W9-CZY-839

INSTRUCTOR'S MANUAL

J. WAYNE WOOTEN

Pensacola Junior College

TO ACCOMPANY

EIGHTH EDITION

ASTRONOMY

THE EVOLVING UNIVERSE

MICHAEL ZEILIK

The University of New Mexico

John Wiley & Sons, Inc.

New York   Chichester   Brisbane   Toronto   Singapore   Weinheim

Copyright © 1997 by John Wiley & Sons, Inc.

This material may be reproduced for testing or
instructional purposes by people using the text.

ISBN 0-471-17746-6

Printed in the United States of America

10 9 8 7 6 5 4 3 2 1

Printed and bound by Malloy Lithographing, Inc.

Instructor's Manual to accompany

ASTRONOMY: THE EVOLVING UNIVERSE, Eighth Edition

By Michael Zeilik

This Instructor's Manual is designed to accompany the eighth edition of Michael Zeilik's Astronomy: The Evolving Universe. It is organized for each chapter of the text as follows:
* an introductory outline of topics in the chapter
* Teacher's notes by Michael Zeilik on the chapter in general
* Wayne Wooten's notes on each topic as presented in the chapter
* suggested slides and overhead transparencies for each chapter, along with possible videos
* an updated listing of articles on these topics from the latest issues of Astronomy magazine
* a set of brief articles on topics to supplement your discussion of the chapter (time permitting)
* suggested class demonstrations or outside astronomical observations
* answers to the study exercises and problems and activities given by Zeilik in the text

Some of the included material is retained from the Instructor's Manual for the previous edition of this text, prepared by Nebojsa Duric of the University of New Mexico.

This Instructor's Manual was written by Dr. Wayne Wooten of Pensacola Junior College, Pensacola, Florida. He would appreciate any notations of errors or suggestions for better wording or revision for future editions. Contact him c/o Physical Sciences, Pensacola Junior College, 1000 College Boulevard, Pensacola, FL 32504-8998, or call him at (904) 484-1152 (voicemail).

Dr. Wayne Wooten has been an active astronomer since he was hooked on astronomy by the appearance of Comet Mrkos in the western evening sky in August 1957; he was a nine year old farm boy at the time. His interest in astronomy led him to found the Walton County Astronomy Club and to sponsor the Escambia Amateur Astronomers' Association in later years. He received his B.S. in Physics from the University of Florida in 1970, graduating Phi Beta Kappa. While teaching science at Paxton High School, he received his Masters in Science Education from the University of Florida in 1972; and after going to work at Pensacola Junior College in 1974, he received his Doctorate in Astronomy Education from the University of Florida in 1979. He has since been named Teacher of the Year at PJC (1986) and the University of West Florida (1991) for his teaching of astronomy at all levels. He also serves as the Staff Astronomer for PJC's Science and Space Theatre.

In 1980 he married Merry Edenton, and they have two boys, Michael (fifteen years old) and Trevor (nine years old). Merry served a six year stint as the Executive Secretary of the Astronomical League (1986-1992). Both have had articles and photos published in Astronomy and Sky and Telescope, and also recently on the "Eyes to the Skies" astrophotography CD-ROM. Michael's observations of the impacts of Comet SL-9 on Jupiter in July 1994 won him Grand Award in Physical Sciences at the Regional Science Fair, and seventh place nationally in the Astronomical League's National Junior Astronomer Award in 1996. Trevor can already identify several constellations and has built his own small telescope, and is especially good at catching meteors and satellites at club and public stargazes. The whole family is active in Boy Scouts, and have helped several Troops with the Astronomy and Space Exploration Merit Badges for Boy Scouts.

By all means, keep looking up!
--Dr. Wayne John Wooten

CONTENTS

CHAPTER 1

From Chaos to Cosmos

CHAPTER OUTLINE

Central Question: What astronomical objects can you see without a telescope, and how do they change with time?

1.1   The Visible Sky
     A. Constellations
     B. Angular measurement
     C. Motions of the stars

1.2   The Motions of the Sun
     A. Motions Relative to the Horizon
     B. Motions Relative to the Stars
     C. Precession of the Equinoxes

1.3   The Motions of the Moon

1.4   The Motions of the Planets
     A. Retrograde Motion
     B. Elongations, Conjunctions, and Oppositions
     C. Relative Distances of the Planets

1.5   Eclipses of the Sun and Moon

1.6   Prehistoric Astronomy
     A. Sunwatching of the Southwestern Pueblos
     B. Anasazi Sunwatching: Ancient Pueblo Calendars

Enrichment Focus 1.1: Angles in Astronomy

CHAPTER OVERVIEW
Teacher's Notes

Since unaided eye observations set the development of our concepts of the universe, encourage (or require) students to make some of the observations described in this chapter. Because of the limited time in a semester, the westward motion of all objects each day (particularly the stars at night) and the eastward motion of the moon are easiest to observe.

See Enrichment Focus 1.1 for specific instructions on observations angles in the sky. Occasionally a faster moving planet overtakes a slower one, and for a few days they appear very close together in the sky. This situation makes the relative motion of the planets most visible. Check the current Abrams Sky Calendars or current copies of Sky and Telescope or Astronomy for conjunctions between planets and stars this semester.

Either the direct or retrograde motion of Mercury and Venus is difficult to observe with respect to the stars because they are usually in twilight or dawn skies, when the constellations are not easily discernible. However, their motions with respect to the sun is very apparent.

Visit a planetarium if one is available to you. Use the planetarium to make the observations and to allow the students to visualize the motions and cycles in the heavens. Don't attempt interpretation of the observations at this point; that develops in chapters 2 through 4. Some students will probably leap to heliocentric conclusions; try to hold them off for a bit.

The students will have a tendency to rely on prior knowledge, particularly of the structure within the solar system, but that knowledge most likely will not include experiencing the retrograde motion of the planets. Any exposure to that motion in a concrete way, through direct observations, plots with the Abrams Sky Calendars, or watching it speeded up in the planetarium, will probably be surprising and puzzling to many of them. Students have many misconceptions about the sky, such as the moon never being visible in the daytime, or that Polaris is the brightest star in the sky.

The eastward motion of the sun or moon, or particularly the planets, will be difficult to visualize or to separate from the daily westward motion of the objects. An analogy, such as a person walking down on a upward-moving escalator, might be helpful. The conceptual problem is that many novice students cannot visualize two motions in opposite directions taking place at the same time. Remind them that all motions are relative!

1.1 The Visible Sky

The motions observed in the sky relate to cycles of time. The students should realize that practically every unit of time we use today was derived first from observations of celestial phenomena. For instance, the day is governed by the sun's transit of the central meridian; A.M. means "ante meridian" (before the local noon passage of the sun through our local meridian); P.M. is naturally "post meridian". We use so much astronomy in our daily lives, without realizing how it influences our culture.

As an example, consider the names of the days of the week. Saturn's day, sun's days, and moon's day are obvious in English, but the other days have Norse instead of Roman gods; in Romance languages like Spanish, the planetary connection is more obvious; Martes is for Mars, Miercoles is from Mercury, Jueves is from Jove (Jupiter), and Veneris is after Venus. Even the choice of seven days was dictated in part by the observation of only seven moving bodies.

The stars are all arranged in constellations and their positions are fixed in (our perspective of) time, and can be precisely measured. The key point is that the constellation figures appear much the same tonight as they did to the Greek astronomer Hipparchus when he first laid out a systematic star catalog about 200 BC. The stars are so distant that to naked eye observation, they are fixed in space in patterns which return every season in the annual cycle.

Students should consider why there are 360 degrees in a full circle; to what time unit does this relate? Most students realize that it is a close and practical approximation for 365.25 days a year. Everyone on earth can see 180 degrees of sky above his horizon at any moment. Of course, your latitude determines the part of the sky you can observe; only at the equator is the entire sky visible.

To show students the daily change due to our rotation, have them observe a bright star low in the west at twilight (Abrams Sky Calendars can be substituted for day classes) and note its setting in an hour. Compare its daily motion with Polaris and with a star rising in the east.

The observed positions of stars over time depends on their positions with respect to the earth, sun, and observer. Stars above the earth's equator, for instance, appear to move most rapidly across the sky (at 15 degrees per hour), while closer to the celestial poles, this motion is not so observable.

In fact, Polaris, very close to the north celestial pole, appears not to move at all in the course of the night. It sits above your northern horizon at an altitude in degrees equal to your latitude in the northern hemisphere. In the northern United States, Polaris sits higher in the sky, but observers there will see less of the southern sky. For instance, in Pensacola, Florida, at 30 degrees latitude, every star within 30 degrees of Polaris is always visible, but any star within 30 degrees of the south celestial pole will never be visible.

1.2   The Motions of the Sun

The daily motion of the sun is defined by local noon, when the Sun is highest in the sky (crossing the local meridian, as in A.M. & P.M.). This is the source of our 24 hour day.

But the sun does not follow the same path daily, varying its noon altitude by 47 degrees over the course of the year. The sun's rising and setting positions along the horizon also vary with this annual cycle, giving us our seasons.

The sun rises and sets farthest north in June (at the summer solstice), and farthest south in December (winter solstice). The days are longest in June, and shortest in December.

Relative to the stars, the sun follows the same path (the ecliptic) through the stars every year, and will be at about the same place on the ecliptic on the same date every year. The 12 constellations through which the sun passes are our signs of the zodiac. The sun appears to move about one degree per day eastward through the stars; remind students this is not an accident; the calendar was rigged that way; 360 is an approximation for 365.25.

1.3   The Motions of the Moon

Like the sun, the moon also appears to move eastward among the zodiacal constellations, but much faster; about 12 or 13 degrees per day or its own half degree diameter every hour (probably the origin of that unit of time). To help students remember the Moon's speed, remind them there are twelve months in the year, so the moon must move about 12 times faster than the

Sun for this to be true. Also note that our week comes from the moon's motion as well; it is the time it takes for the Moon to complete each quarter phase (new moon to first quarter).

Use the Abrams Sky Calendars to note the moon's phases and its passage by bright stars on the ecliptic. The students can even use calendars over a two month interval to derive the length of the sidereal month (star based, 27.3 days) and synodic month (phase-based, about 29.5 days).

1.4   The Motions of the Planets

While the planets are so distant they do not show visible disks with the naked eye, they reveal their nonstellar identity by moving among the stars through the constellations of the zodiac.

Like all distant objects, the planets seem to rise in the east and set in the west each night as the earth spins. But as we compare their positions night after night, we observe that most of the time the planets move eastward (direct motion) in the same direction as the sun and moon move. They move in a wide variety of speeds, however, with Mercury moving fastest; Pluto the slowest. To further complicate matters, in intervals of several months the planets (*wanderers* in Greek) will stop moving eastward and instead retrograde westward.

The retrograde motion has two patterns. The inferior planets, Mercury and Venus, have this occur just after greatest eastern elongation, when they are farthest from the sun in the evening sky. For the superior planets, retrograde occurs near opposition, when they rise in the east at sunset.

Abrams Sky Calendars can, over a period of two months, be used to show students the motions of most bright planets. In particular, Mercury can be shown to retrograde in either evening or morning skies in just a few weeks. Even Jupiter and Saturn will shift somewhat among the stars during the course of most months.

1.5   Eclipses of the Sun and Moon

Solar eclipses occur when the new moon passes between the sun and the earth, covering at least part of the sun behind the moon's disk. As the moon's orbit is tilted five degrees with respect to the ecliptic, only about one in six new moons occur close enough to the ecliptic to produce a solar eclipse. Remind the students that both the sun and moon appear a half degree across, but the moon can ride as much as ten times its diameter above and below the ecliptic, so the Moon's shadow will pass well above or below the earth at most new moons.

It is worth stressing to the students just how deceptive sizes in the sky can be. The Moon is the same apparent size as the sun, both about a half degree across, yet the sun in truth is four hundred times larger in diameter (and thus almost four hundred times more distant). Get the students to relate these distances to the apparent motions (closer objects move faster).

Twice a year, at "eclipse seasons", the moon does cross the ecliptic at either new or full phase, and someone on earth will see an eclipse. In fact, you may get both a lunar and solar eclipse in a season, but opposite sides of earth will see them, occurring about two weeks apart.

The author does not distinguish between total and annular solar eclipses, but this was important on May 10, 1994. The moon was far enough away from Earth on that day that its disk was considerably smaller than the sun's; even though observers in the United States on a line running from El Paso, Texas northeast through most of New England saw the new moon on the

4

ecliptic passing directly in front of the Sun, there was no totality. Instead, there was a bright (and DANGEROUS) ring of solar fire still visible at even the center of the eclipse. Everyone in the United States with clear skies saw a partial solar eclipse on that day, but in all cases, safety precautions normally used in observing the Sun should have been stressed.

Only during TOTALITY can an observer safely stare at the totally eclipsed Sun and witness the corona and prominences with the unshielded naked eye. As the U.S. does not have such a total eclipse until August 21, 2017, make sure the students understand this.

Lunar eclipses are much easier to observe, as anyone on the night side of the earth can look up and see the earth's shadow projected upon the disk of the Full Moon as it passes through the ecliptic. This shadow's color depends greatly on the state of our atmosphere during the eclipse. If the stratosphere is very dusty (such as the volcanic eruption in the Philippines in 1991 and 1992), then only the very longest wavelengths can be bent around the rim of the Earth and reach the moon during the eclipse; hence the moon in total eclipse may appear bright orange to deep red. From my personal observations, the eclipse of the moon in May 1975 was a beautiful peachy color, but the two lunar eclipses of 1982 were both very dark, due to large volcanic eruptions several months before these eclipses. Obviously our atmosphere was much clearer in the 1975 eclipse than it was in 1982. Photos of the eclipse on December 1, 1992 showed that eclipse was very dark, at times invisible with the naked eye during totality. Ask some of your students if they remember seeing it. Compare it with the much brighter lunar eclipse of November 28-29, 1993.

1.6   Prehistoric Astronomy

In Cosmos episode 103, Sagan visits an Anasazi site, illustrating the calendar quite graphically. In his Astronomical Calendar, Guy Ottewell names the months by their native American titles. Those students with native American ancestors will find including such material makes the course more enjoyable for them.

ADDITIONAL RESOURCES

John Wiley and Sons, the text's publisher, provides instructors with sets of unique slides to supplement this text. Listings of them will be given for each chapter, along with acetate overhead transparencies for that chapter. For chapter 1, the slides are the annular eclipse opening the chapter, figures 1.12ab, 1.18, 1.24a, and 1.27. Transparencies include diagrams 1.2, 1.3, 1.6ab, 1.10abc, 1.11, 1.14, 1.16, 1.19, 1.20, 1.21ab, 1.22, 1.25, and 1.26.

I highly recommend Abrams Planetarium Sky Calendars for the current months as class handouts. A subscription is only six dollars per year, available from Sky Calendar, Abrams Planetarium, Michigan State University, East Lansing, MI 48824. Write them for permission to reproduce up to four months for classroom use, or inquire about bulk discounts for originals for all your students. We will have opportunities to use these with several chapters.

For the instructor's reference, The Astronomical Calendar for the current year is another invaluable aid in noting various planetary motions over the entire year. The notes on eclipses include excellent diagrams and observing tips. It is available toward the end of November of the preceding year for $15, and can be obtained from the Astronomical Workshop, Furman University, Greenville SC 29613; call (803) 294-2208 and get put on the permanent mailing list.

The workshop also has other highly recommended publications, including the Astronomical Companion ($12), a well illustrated cosmic zoom, and The Under-Standing of Eclipses ($12.95).

To stay current, both Astronomy and Sky & Telescope magazines should be read by astronomy instructors. For the students at the beginning level, Astronomy is probably the better choice; Sky & Telescope has good historical background, but is more technical, in general.

To that end, I will include with each chapter a list of pertinent articles from the last several years of Astronomy to supplement that chapter. In some cases, I will also refer to especially good articles in later Sky & Telescope issues in the additional discussion topics. Contact Astronomy by calling 1-800-446-5489 or writing PO Box 1612, Waukesha, WI 53187; it costs $24 yearly. Sky & Telescope is $30/year; write Sky Publishing at 49 Bay State Road, Cambridge MA 02138, or call (617) 864-7360. Both magazines have numerous other useful publications, and will forward free catalogs upon request.

With many students having access to computers at home or school, there are a good variety of planetarium programs which can be used for student exercises. My personal favorites are Skyglobe, Skymap, and Dance of the Planets. The first two are shareware, and can be downloaded off the Internet or from many local BBS services. They are also inexpensive to register, with Skyglobe just costing $25, and Skymap only $40. Dance is more sophisticated, with great graphics, and a good value for $100; a limited feature demo version is also available on many BBS services.

I like Redshift's graphics a lot, but it does not allow the inputting of new discoveries, as do Skymap (a snap) or Dance (tricky); I will discuss how to add new comets like Hyakutake in chapter 12.

The May 1994 issue of Astronomy and September 1996 issue of Sky & Telescope have good directories of such programs available on both PC and Macintosh formats. Encourage your students to at least get the shareware versions to use during the term. Computer users are also increasingly finding a lot of astronomical discoveries on the Internet; Sky & Telescope lists a lot of great sites in the August 1995 issue on page 20, as does Astronomy in June 1995 on page 74. By all means, start with the Sky and Telescope's own award winning web site at http://www.skypub.com. Their weekly astronomy news bulletin is an award-winner, as are their comet bulletins! Astronomy magazine's home page is at astronomy.com; both sites branch off to a lot of other great hits as well.

SUPPLEMENTARY ARTICLES
1. "Clocks and Calendars in Astronomy," Astronomy, January 1984, page 51.
2. "Computing Time of Sunrise and Sunset," Astronomy, April 1984, page 75.
3. "Indian Observatories of Jai Singh," Astronomy, January 1985, page 18.
4. "Sundials and How to Build Them, " Astronomy, January 1986, page 47
5. "Eclipse Predictions for your Computer," Astronomy, November 1986, page 67.
6. "Watching the Earth Rotate with a Shadow Clock," Astronomy, August 1987, page 32.
7. "A New Slant on Earth - Climate and Axial Tilt," Astronomy, July 1992, page 44.
8. "Capture a Constellation," Astronomy, November 1992, page 66.
9. "Improving Astronomy Education in America," Astronomy, September 1993, page 40.
10. "Charting a path in the Night Sky," Astronomy, October 1993, page 74.

11. "Return to Darkness: Nov. 1994 Totality," <u>Astronomy</u>, November 1993, page 88.
12 "Astronomy Software Buyer's Guide," <u>Astronomy</u>, May 1994, page 55.
13. "Not Just Another Pretty Face (Lunar Phases)," <u>Astronomy</u>, July 1994, page 76.
14. "Under the Southern Sky," <u>Astronomy</u>, October 1994, page 72.
15. "Blasting Along the Infobahn (Internet Astronomy)," <u>Astronomy</u>, June 1995, page 74.
16. "Students in Cyberspace," <u>Astronomy</u>, October 1995, page 48.
17. "Tilt-A-Whirl Astronomy," <u>Astronomy</u>, March 1996, page 50.
18. "An Exotic Eclipse," <u>Astronomy</u>, April 1996, page 74.

ADDITIONAL DISCUSSION TOPICS

Eclipse and Solar Observing Safety

Eclipse safety is worth stressing. The sun does not produce any unusual forms of energy just for the eclipse; it certainly does not "know" the moon is coming between itself and the earth. But one can easily become partially blinded by staring at the sun on any clear day; only on eclipse days are some people foolish enough to try. As a solar eclipse approaches, look for adds for aluminized mylar from several companies to allow safe viewing of the eclipse. Such viewers for your eyes or binoculars will probably only cost $1-2, and might even be a good investment for your whole class. They will be vital for all solar observing and can be used to observe the Sun safely for years to come.

Be sure students understand that the filters must cover the front lens of their telescope or binoculars. Eyepiece filters are NOT recommended for any type of solar viewing, as the Sun's intense heat can melt or crack them, and anyone buying a telescope with such a filter screwing onto the eyepiece should probably just discard it.

If you look in <u>Astronomy</u> or <u>Sky & Telescope</u>, you will also find filters designed to put over the front aperture of your telescope. They are expensive, but allow safe visual (and photographic) observation through the eyepiece with a more natural orange or yellow solar image. Thousand Oaks Optical produces them in a variety of sizes. Write them for current prices at Box 5044-289, Dept. A, Thousand Oaks, CA 91359, or call (805) 491-3642. The 2.4" aperture (unmounted) for $39 is a best buy, for it transits plenty of light of even an 8" scope. Mount the filter on an off-axis mask.

Eclipses of note

While there are no more total or annular solar eclipses in the US until 2000, students might like to observe several other eclipses coming up in the rest of the century, provided they are willing to travel. After all, the umbral shadow is only about a hundred miles wide under the best conditions.

The August 1996 issue of Sky & Telescope has a good article on upcoming solar eclipses for the rest of the century on page 48. The eclipse in March 1997 in Siberia and Outer Mongolia will be very cold, but promises a naked eye view of Comet Hale-Bopp during totality, a very rare celestial event. Only twice before in human history do we have records of a naked eye comet being seen during a total solar eclipse--be sure to pack your long johns if you are game!

Latin America fans will find the total eclipse of February 26, 1998 ideal. This long totality runs through northern Ecuador into Venezuela, then through several islands of the southern Caribbean; Aruba and Curacao are very nicely placed. Watch science magazines for tours and cruises already being organized for this one. This would be a great trip to get to know the southern skies we never see from most of the northern hemisphere as well, and the best eclipse anywhere close to the U.S. in the rest of this century.

Those desiring a European vacation might schedule it around August 11, 1999, when the total eclipse shadow runs from southern England, then just north of Paris, through downtown Munich, and down through the Balkans into Turkey and India.

Lunar eclipses require much less traveling, as they are visible from the entire night side of the Earth. Total lunar eclipses visible from the United States are sparse now, the next is January 21, 2000. Partial lunar eclipses in this time period include March 24, 1997 (93% eclipsed), and July 28, 1999 (42% eclipsed). Check out articles in astronomy magazines for more details.

DEMONSTRATIONS

Shadows and the sizes of the Sun and Moon

How do we know the Sun is more distant than the Moon? First, the Sun appears to move slower (one degree per day eastward) than the Moon (twelve degrees per day). Secondly, in a solar eclipse, the moon's black disk is seen silhouetted in front of the sun. If it is a total solar eclipse, the sizes match and the moon fits exactly in front of the sun's photosphere; the fortunate few on Earth to see totality lie in the moon's umbral shadow. For the annular eclipse of May 10, 1994, the moon was little too distant from earth to be large enough to completely block the sun, thus a ring ("annulus" in Latin) of light surrounded the moon. As most of us were a few hundred miles above or below the central line of the eclipse, in the moon's penumbra, then we saw just a partial eclipse, with only part of the Sun visible. All of the continental United States observed the partial phases of this eclipse.

This next demonstration will allow students to visualize how the relative sizes of the sun and moon were first established. Bring a basketball and a tennis ball to class. Have two students measure the sizes of both balls, and write the sizes on the board.

Designate one student as the observer, and let the two students with the balls line them up so that the closer tennis ball (the moon) appears to fit exactly in front of the larger but more distant basketball (the sun).

Now measure the distances from the observer to each ball, and write these distances on the board. Find the ratios of both the balls' sizes and distances. It is obvious that if the larger ball is eight times bigger, it had to be placed eight times farther away to appear the same size as the closer ball.

It was this line of reasoning that allowed the Greek astronomer Aristarchus to first estimate the Sun's size. Aristotle had shown from our own shadow on the moon at lunar eclipses that the moon was only about one quarter the size of Earth. Aristotle also correctly noted that the earth is a sphere, for it always casted a circular shadow at every lunar eclipse.

From his observations of the times of first and third quarter moons, Aristarchus deduced the sun must be about twenty times farther from us than the moon, so must also be twenty times larger than the earth's satellite. If so, then the sun was at least five times larger than earth. Since the smaller moon orbited the larger earth, Aristarchus reasoned that the smaller earth should orbit the still larger sun, the first statement of the heliocentric model known.

ANSWERS TO STUDY EXERCISES

1. The ecliptic can be traced roughly by the positions of the moon and planets visible in the night sky. Note the sunset point on the western horizon, then trace out an arc through the visible planets and moon. Those constellations lying along this arc are the signs of the zodiac; many of them are bright enough to be easily identified by their star patterns, such as Scorpius, Sagittarius, Taurus, Gemini, and Leo. The ecliptic passes through all twelve zodiacal constellations.

2. Refer to figure 1.21, being sure the directions east and west are correct.

3. The sun, moon, and stars never retrograde; only planets can retrograde.

4. The inferior planets, Mercury and Venus, retrograde when near inferior conjunction, passing in front of the sun. The superior planets, Mars, Jupiter, and Saturn, will retrograde near opposition, rising in the east at sunset.

5. At opposition, a planet rises at sunset, and sets at sunrise (opposite the sun in the sky).

6. The moon moves faster than the sun in the sky (twelve degrees per day, as opposed to one degree per day for the sun); as a rule, closer objects move more notably. At a solar eclipse, it is obviously the dark moon passing between us and the more distant brilliant sun.

7. The waxing crescent moon is setting; the following evening, the moon is about twelve degrees higher and farther east; it is waxing, and a thicker crescent than the night before.

8. At equinoxes, the sun rises due east, and sets due west. At the Vernal Equinox (March 21) the sun rises farther north each morning, reaching its northernmost point at summer solstice about June 21. It then turns back south, and is farthest south on December 21 at winter solstice.

9. Solar eclipses require a new moon between us and the sun; lunar eclipses happen when the full moon moves into our shadow.

10. Mars takes only about two years to circle the ecliptic, while Saturn takes about thirty years; as a rule, faster moving bodies are closer to us.

11. Before the solstice, note the sunrise position on the SE horizon with a marker. Each clear morning, watch to sun to set farther south until solstice, then turn back toward your marker after solstice. Note the date and number of days until it again rises at your horizon marker; the solstice date is exactly between the marker dates.

12. Due to the tilt of the ecliptic, the angular speed along the horizon is greatest when the sun is rapidly climbing (Vernal Equinox) or falling (Autumnal Equinox). The sun's path levels out and it moves almost due east at the solstices for several weeks; then its horizon point changes little.

13. From winter to summer solstice, the sunset point moves northward daily.

14. At a mid-northern latitude, the noon sun is NEVER overhead. Due to the 23.5 degree tilt of the earth, the maximum northward elevation of the sun is on June 21, passing through the zenith for observers at the Tropic of Cancer (+23.5 degrees latitude). Only if you live in the tropics (between +23.5 and -23.5 degrees) can you observe the noon sun overhead.

15. A week after the Vernal Equinox, the sun will be higher in the sky, thus the shadow of the pole will be shorter; if the sun is overhead at your zenith, no shadows are cast by the pole at all.

ANSWERS TO PROBLEMS & ACTIVITIES

1. The moon moves 360 degrees in 27.3 days, or 13.2 degrees per day, or .55 degree per hour; as the moon is about a half degree across, the moon basically moves its own diameter every hour.

2. The sun's is about one-half degree across, so the small angle equation gives sin .5 deg = .00873, or the sun distance to the sun is 1/.00873 = 114.6x greater than the solar diameter. To check, the sun's distance is one A.U., or 1.496×10^8 km distant, and it is 1.39×10 km in diameter. Thus the sun is 107.6x smaller than the A.U., and 107.6 solar diameters separate us from it, thankfully.

3. The moon's disk is also about one-half degree across, so we get the same size to distance ratio of 114.6 as we got in # 2 above. The moon is 3,476 km in diameter, and orbits us at an average distance of 384,000 km; this it is 110.5x smaller than its average distance from earth. The sun is 107.6x smaller than its diameter, so both appear about the same size in the sky (about one-half degree across), although the sun is really 389.6x more distant, and 399.9x larger than the moon.

4. Tan h = 10/3.5 = 2.857, thus the altitude of the sun is 70.7 degrees.

5. In a month, the sun moves about 360/12 = 30 degrees, or about a degree per day. As there are sixty arc minutes per degree, the sun moves 60/24 = 2.5 minutes per hour. As there are twelve months in the year, the moon must be moving about twelve times faster than the sun; as 12 x 2.5 = 30 arc minutes, or one-half degree, this again shows the moon moves its own apparent diameter eastward in the sky every hour.

6. The moon's disk appears 3 mm in diameter, so the scale is 6 mm per degree. The moon appeared about one degree from Venus in the first exposure, and only a half degree from the moon in the last exposure; the moon moved about a half degree, or its own diameter during this sequence. Incidentally, the moon and Venus appeared to move 9.6 mm, or about sixteen degrees during the sequence. As the earth rotates at fifteen degree per hour, the exposure took a little more than an hour, the approximate time it takes the moon to move its own diameter.

CHAPTER 2

The Birth of Cosmological Models

CHAPTER OUTLINE

Central Question: What is a scientific model, and how did early models explain and predict astronomical observations?

CHAPTER OVERVIEW
Teacher's Notes

       In diagraming and explaining the Ptolemaic model, neglect at first the daily motion (as the text has done). That will cause less confusion. This may be added at the end as a rotation of the entire system about the earth once each day. The sphere of fixed stars would then turn at a rate different than that of the sun, one day more each year. Do not let the students get hung up on the details of the equant! It is sufficient that it is a violation of the precept of uniform motion about the center of a circle but needed by Ptolemy to account for certain observations.

11

Again, because of prior knowledge, students will have a hard time visualizing the Ptolemaic model as worthy of consideration. If asked to devise a model to explain the observations in chapter 1, most will chose to start with a heliocentric one, even though to an earth-bound observer in the sixteenth century, a geocentric model is more natural. It wouldn't hurt to emphasize at this point that there is no unaided -eye evidence to support a moving earth. you might propose a geocentric model and then challenge the students to refute it, using only the evidence presented so far. We have used a geocentric/heliocentric "shoot out," where the teacher defends the geocentric model and a student tried to fend for a heliocentric model. The arguments must always go back to just the naked eye observations of Chapter 1.

This chapter introduced the idea of a model, an idea that is basic to modern scientific thought and that will appear again and again throughout this text. There are different kinds of models, quantitative and qualitative, but basically models are mental pictures and predict the behavior of that system. To get across to students the function of models in scientific inquiry, use examples, such as our description of an atom. Ask students to describe a hydrogen atom (most will have some idea of the Bohr model), and point out that they are describing something which has never been seen (but probed, nonetheless) that a given description or model predicts behavior which may not have been studied, and that a real atom does not consist of a spherical electron orbiting around a spherical proton.

The predictive power of a model may be illustrated as follows. If someone describes an object as being made of steel, than that model predicts a specific interaction between the object and a magnet, which may be tested. Failure of the object to behave as predicted will require a modification of the model! If you try this, it will take a lot of will power not to tell the students what objects really are. point out that in many cases, we will never really see or know the real system; we will have only models to work with!

2.1   Scientific Models

Stress that models are not intended to be perfect; a good scientist should not become overly upset if new data fails to support his model. First the model must explain observations already made, then it should predict new observations. The classic case is Einstein's Theories of Special and General Relativity; they explained the previously noted anomalies in the motion of Mercury, then predicted the 1919 gravitational lensing of starlight by the Sun during the total solar eclipse. But Einstein knew that there was much more; on his deathbed was work on the Grand Unified Field Theory, which is still far from complete. All models are tentative, the best we can do under present conditions.

2.2   Babylonian Skywatching

The names of the brightest constellations are chiefly derived from Babylonian and other middle eastern cultures. They noted early on the five naked eye planets, and gave them divine powers, leading in part to the development of astrology. Their careful observation of the moon's motion lead them to discovery of the Saros cycle of eclipses. With their growing powers of predicting future celestial events, they were giving birth to modern science.

2.3   Greek Models of the Cosmos

In the progression of the geocentric models, note that Aristotle proposes the basic model about 340 BC. Hipparchus adds the eccentrics, epicycles, and deferents about 130 BC, and Ptolemy adds the equant and fine tunes the model around 130 AD. Many times in science we will see initial models constantly revised as better data becomes available. Certainly Hipparchus and Ptolemy were among the finest observers of the pretelescopic era, and they were challenged by the accuracy to their new observations to develop a model which could match their past observations and predict future planetary positions with great accuracy. It is a real tribute to the work of Ptolemy that he reworked the geocentric model to the point that it predicted the positions of the planets with (for naked eye observers) reasonable accuracy for the next 1,400 years.

Yet the correct heliocentric model had already been proposed by Aristarchus even before Hipparchus devised his geocentric revision. His model had the Earth rotating on its axis once a day, and revolving around the Sun once a year. Why did the Greeks not accept the first incarnation of the Sun centered model?

Greek physics required that all heavenly bodies moved on actual physical spheres, hence accounting for the heavenly motions. The motion of Earth would crash through these divine spheres. Also if we were revolving, over a six month interval we should be able to observe stellar parallax, due to our changing vantage point with respect to the celestial sphere. Of course we know today this is observed, and in fact used to find the distances to at least the closest of the stars. In ancient times, the concept of a cosmos so vast (the closest stars are hundreds of thousands of astronomical units distant) was beyond their imagination. Consider that M (1,000) was the largest Roman numeral.

2.4   Claudius Ptolemy: A Total Geocentric Model

Note how thorough Ptolemy is. He tries to find a cause for every observed motion, even minor variations in planet motion and brightness from opposition to opposition. The man would have made a great researcher and theoretician, regardless of when he was born! Note in particular how he used the eccentric offset of the earth and the equant to explain these variations. Think of the math and geometry involved in finding the best fit to the observational data.

When Copernicus revived the heliocentric model in the next chapter, he was not nearly as careful as Ptolemy; in fact, his model did not predict the planetary behavior any better than did Ptolemy's model. It was Johannes Kepler who first matched the great mind and striving for perfection which so drove Claudius Ptolemy. No wonder the Arabs regarded his work as "the greatest"--Alamgest.

ADDITIONAL RESOURCES

Transparencies 2.1, 2.4, 2.5ab, 2.6, 2.9ab, 2.11ab, and F1 will supplement your lecture.

The intellectual debate between geocentric and heliocentric models of the universe is illustrated extremely well in the third episode of Carl Sagan's television classic, "Cosmos".
It is now available on VHS, and probably can be even rented at a local video store, or purchased through the Astronomical Society of the Pacific. Buy the entire set for about $200, and even though some of the episodes on planetary exploration have become dated, several of the episodes are as good as ever, including this one. Call ASP at (415) 337-2624 for info.

SUPPLEMENTARY ARTICLES

1. "Nightviews--The Philosophy of the Astronomer," Astronomy, August 1991, page 34.
2. "The Grand Illusion," Astronomy, November 1992, page 44.

ADDITIONAL DISCUSSION TOPICS

A Tribute to Early Astrologers

Realize that astrology began as observation; the position of the sun in the sky determined when we planted and harvested our crops. As fisherfolk, our activities were governed by the moon and the tides it propelled. Did it not make sense, therefore, that the five other bodies moving among the constellations must also influence our lives. But these influences must be more subtle, for the planetary cycles could span generations. This drove the astrologers to compile accurate records over periods of many centuries, records we still use today in tracing back "broom stars" (comets) and "guest stars" (novas and supernovas) back thousands of years into the past. Of course, the chinese and indian astronomers were also compiling similar records, but the Greeks utilized the Babylonian works far more in their development of cosmology.

DEMONSTRATIONS

Dance of the Epicycles

Have a student, the earth-based observer, sit in a large open area in a chair. Draw a circle about 20 feet in radius around our observer. Then have a pair of students to model Ptolemy's geocentric model; one, the deferent for the planet, will walk exactly along the circle counter-clockwise, at the same time swinging his partner around slowly on her epicycle. Have the central observer call out when the epicycling dancer seems to stop moving counterclockwise and starts retrograding. Note this is always when she is on the inside of the circle, closest to our earth-based observer.

Looking ahead to Chapter 3, the dance will now require the earth-based observer to become a participant. Now in a heliocentric model the student representing the sun gets to sit in the center, with a fast student (Mercury or Venus) moving around the sun at a faster pace then our earth observer walking around the same circle. The earth observer notes the faster student lapping him as the faster student passing between him and the sun (inferior conjunction). An even more distant student walking slower than our earth observer will seem to retrograde from our point of view when we overtake him at opposition, in accord with chapter 1.

If you have a planetarium with a Digistar projector, the epicycles show up well by turning on the "trail on" feature while running time forward for several months around the time of opposition for a superior planet, or inferior conjunction for Mercury or Venus. Many computer programs also have a planet tracking option to allow retrograde motions to be traced out and even printed out for the students. On Skymap, for instance, find the planet, click on the right mouse button, and you will get a menu of how to track a planet on your chart for several months, including retrograde loops.

ANSWERS TO STUDY EXERCISES

1. The Babylonians made accurate and careful observations of the motions of the planets, originally named many of the modern constellations, found variations in celestial cycles, and developed numerical or arithmetic methods of predicting planetary motions and motions, as well as eclipses.

2. The Greeks paid little attention to careful observations, compared to the Babylonians, but they did devise geometrical models to explain astronomical observations.

3. In Aristotle's model:
a. Natural motion of earthly material is to fall toward the center of the universe. Once there, it has no natural motion.
b. The natural motion of the celestial sphere is to rotate. The sphere of the fixed stars did so daily from east to west.
c. The sun's motion was again the natural motion of that celestial body's sphere yearly from west to east. Evidence that the sun moved and not the earth was lack of an annual parallax of the stars.

4. The spherical shape of the earth assumed a spherical geometry dominated the entire universe, aesthetically a pleasing conclusion. Physically, the earth is made of material whose natural motion was toward the center of the cosmos. Also, the lack of an annual stellar parallax implied that the earth did not move around the sun. Note that parallax here is NOT the same as the modern stellar parallax used to measure the distances to the closer stars.

5. According to Ptolemy, epicycles produce retrograde motions. Thus the time between the beginning of the retrograde motions, essentially the synodic period, must be the epicycle's period. The motion along the deferent produces the normal west to east (eastward) motion of the planet along the ecliptic. So the period of the deferent is the length of time it takes the planet to circuit the zodiac once (twelve years, in the case of Jupiter).

6. Mercury and Venus are seen in the general direction of the sun when undergoing retrograde motion, and are never seen at opposition. So Ptolemy placed the deferents for Mercury and Venus between the earth and the sun, to explain the observed phenomena. The size of the epicycles were set by the maximum elongations of Mercury and Venus from the sun, so Venus has the larger epicycle.

7. A basic assumption of the Ptolemaic model was that the natural motion of celestial bodies was uniform along circles. The equant produces nonuniform speeds along the deferent, so uniform motion was then observed from a point off center of the deferent.

8. Ptolemy said that:
a. the epicycle produces retrograde motion.
b. the eccentric produces nonuniform eastward planetary motion along the ecliptic as seen by us.
c. the equant produces variations in retrograde motion.

9. It was in harmony with the physics of the day, especially the concepts of natural and forced motion.

10. In the geocentric model, NO annual stellar parallax occurs.

11. A heliocentric model predicts that annual stellar parallax should occur.

12. Geometry outlines a visual framework for the model.

ANSWERS TO PROBLEMS & ACTIVITIES

1. $8/360 = .022$ of the earth's full circumference. So the circumference would be 360/8 x 5000 stadia, or 45 x 5000 = 225,000 stadia. Then the radius is 35,800/2x3.1416 or about 6,000 km.

2. One degree contains 60 arc minutes. The earth-sun distance is one AU. So the stars would have to be placed at least 60 astronomical units for the parallax to be too small to observe with the naked eye.

3. The offset would need to be $(186-179)/(186+179) = 1.9\%$ of the diameter of the circle; for a 10 cm radius, this would be 3.8mm from the center.

4. When the earth is closest to them, the stars are only 9 AU from the sun, and are thus separated by $X = \sin 6$ deg. x 9 AU = .94 AU. When the earth is on the other side of its orbit, at 11 AU, the angle would be $\sin X = .94/11 = .085$, so the angle would shrink to 4.9 deg., a change of slightly over a degree, or about two moon diameters, easily seen with the naked eye.

5. The moon is about 110x its own diameter distant from us, and about a quarter as large as the earth. As the earth is about 12,800 km in diameter, the moon is close to 3,200 km across, or about 110 x 3,200 = 350,000 km distant. Thus the circumference of the moon's orbit is about 2.2 million kilometers, and the moon revolves around it in 27.3 days, moving 81,000 km per day, or 3,400 km per hour. This is in good accord with the observation that the moon does appear to move about its own diameter eastward across the sky every hour.

6. If the moon is 1/4 earth diameter, it must be about 350,000 km from earth on average.

7. The sun's diameter is 107x smaller than the AU, then the sun must be 107x larger than the moon and earth, or 342,000 km across. As the sun moves its diameter in half a day, this means the sun is moving at 342,000 km / 12 hrs. = 28,500 km/hour.

CHAPTER 3

The New Cosmic Order

CHAPTER OUTLINE

Central Question: How did Copernicus' sun-centered model explain the motion of the planets and ignite a revolution in cosmological thought?

3.1

       Copernicus the Conservative
       A. The heliocentric concept
       B. The plan of <u>De Revolutionibus</u>

3.2   The Heliocentric Model of Copernicus
       A. Retrograde motion explained naturally
       B. Planetary distances
       C. Relative distances of the planets
       D. Problems with the heliocentric model

3.3   Tycho Brahe: First Master of Astronomical Measurement
       A. The new star of 1572
       B. Tycho's hybrid model

3.4   Johannes Kepler and the Cosmic Harmonies
       A. The harmonies of the spheres

3.5   Kepler's New Astronomy
       A. The battle with Mars
       B. Properties of ellipses
       C. Kepler's Laws of Planetary Motion
       D. The new astronomy

Enrichment Focus 3.1: Sidereal and Synodic Periods in the Copernican Model

Enrichment Focus 3.2: Geometry of Ellipses

CHAPTER OVERVIEW

Teacher's Notes:

Students generally have no trouble visualizing a heliocentric model; it's the one they have lived with for many years! So picturing a heliocentric model may very well be more natural than the geocentric model in earlier chapters. But the point should be made that no naked eye observations, including those of Tycho Brahe, can distinguish between the two models, although such observations can distinguish between circular and elliptical paths for the planets.

Perhaps because of their familiarity with the Copernican model, students may have the notion that the Copernican model is simpler and more accurate than the Ptolemaic once. This is NOT so--the heliocentric model was simply more satisfying conceptually and aesthetically to Copernicus. Tycho's model was perhaps more "natural" at the time, and could easily have been adopted had his observations been analyzed by a believer in a geocentric universe (as Tycho was himself). Perhaps this would be a good place to introduce the role that scientific "taste" plays in the development and acceptance of a model.

A note on Tycho's model: Have students compare this model to the Copernican one. They should see that like Copernicus' model, Tycho's allows for the determination of the relative spacing of the planets. The only change from Tycho's model to that of Copernicus is a change in the reference point from the earth to the sun.

Most student will find concrete demonstrations or examples helpful in picturing synodic and sidereal periods or the determination of the distances to planets. Graphical examples may make a good lab exercise. Have the students determine the distances to Jupiter or Mars by plotting positions of the planets on a heliocentric diagram. (The more math inclined may use trig to calculate these distances). Raw data, such a dates and times of oppositions, conjunction, and quadrature, may be found in Ottewell's Astronomical Calendar.

Along these lines, note that the text illustrates the determination of the distance to Mars using observations over one Martian sidereal period. Using this method for Jupiter would require observations over an 11 year period. The distance measurement can be made over a shorter period of time by observing the planet first at opposition and then at quadrature, about three months later, once this sidereal period is known. Again, begin with raw data and have the students carry our the graphical analysis to find the distance to the planet. The purpose here is not to simply find the distance, but to go through the process in a concrete, hands-on way.

3.1   Copernicus the Conservative

Note how long Copernicus worked on the model; his original model was shown to friends in 1514, but his book was not to appear for thirty years. Yet despite all his work, his final model did not do a better job of explaining planetary behavior than Ptolemy's. Copernicus appealed for a heliocentric model rather on aesthetic grounds; it is thus perhaps it was taken as seriously by other scientists as it was.

Note that like the ancient Greeks, Copernicus thought that God loved circles. A model with geometric symmetry and beauty must be right. This same concept was to start Kepler on his research into planet orbits, as well shown in Cosmos episode 3. For course, Kepler was to find reality very different from the perfect circles of Copernicus.

3.2 The Heliocentric Model of Copernicus

To get a feeling for the intellectual foment his book stirred, consider the title <u>De Revolutionibus</u>; before its publication, revolution referred only to orbit--now it was to visualize a complete rethinking of how things worked, with political as well as scientific implications.

3.3 Tycho Brahe: First Master of Astronomical Measurement

While Zeilik mentions that Tycho was unable to observe any parallax for his supernova in 1572, he does not mention Tycho also exchanged reports on bright comets with other observers. Again, less parallax than the moon's was observed, suggesting that comets were considerably more distant than the moon, at least out in the realm of interplanetary space. This is very different from the Greek view that comets were noxious vapors in the earth's upper atmosphere, much as we think of meteors appearing today.

3.4 Johannes Kepler and the Cosmic Harmonies

Note that Kepler's initial insight, about the five regular solids fitting in the planet orbits, was dead wrong. But it does show the power of his intellect, getting him invited by Tycho to a job that would unite a great stockpile of accurate planetary data with perhaps the one person with the math skill and perseverance to finally derive the laws of planetary motion. When they did emerge, they vastly disappointed Kepler, for his five solids vanished in smoke. But he accepted that he had been wrong, and built upon both his failures and successes. For science to make progress, we must have thinkers willing to be wrong, and from their mistakes to gain insight. The worst crime in science is NOT to think at all--to never tackle the tough problems.

3.5 Kepler's New Astronomy

Note that Kepler's first law does not mean a planet can't have a circular orbit; Venus' orbit is very close, and most are of low eccentricity (Pluto, Mercury, and Mars have the highest eccentricities). A circle is a special ellipse with an eccentricity of 0. At the other extreme, if you took your two focal points and kept separating them still farther, you would end up with a straight piece of string connecting them; then the interfocal distance and the major axis would be identical, and the eccentricity would be 1; this is easy to remember, as a circle looks like a 0, and a straight line the number 1.

With respect to Kepler's second law, note it applies to all planets; Pluto must sweep out the same area in a month as does Mercury. But of course Mercury is only .4 AU from the sun, so it sweeps out 1/3 of its orbit in a month. At 40 AU, Pluto is 100x farther way, so it needs to move 100x slower to still sweep out the same area.

The third law's squares and cubes are tough for students, but an idealized example is an asteroid with an orbital period of 8 years. Squaring 8 gives 64, so the average distance of this asteroid is the cube root of 64, or 4 AU from the sun. Saturn also works well; squaring its period of about 30 years gives 900, or close to 1,000. The cube root of 1,000 is 10 AU, not far off from Saturn's distance as calculated by Copernicus. It is worth noting that Copernicus was farther off on his estimates of Saturn's distance than with any other planet because Saturn moved slowly against the stellar background, so there was little change in its position in the three months between opposition and quadrature.

ADDITIONAL RESOURCES

Transparencies 3.3, 3.4, 3.6, 3.15, 3.16, 3.17, and 3.18 will assist you in class.

The intellectual debate between geocentric and heliocentric models of the universe is illustrated extremely well in the third episode of Carl Sagan's television classic, "Cosmos". It is now available on VHS, and probably can be even rented at a local video store, or purchased through the Astronomical Society of the Pacific. Buy the entire set for about $200, and even though some of the episodes on planetary exploration have become dated, several of the episodes are as good as ever, including this one. Call ASP at (415) 337-2624 for info.

SUPPLEMENTARY ARTICLES
1. "Tycho Brahe lights up the Universe," Astronomy, December 1990, page 28.
2. "Nightviews--The Philosophy of the Astronomer," Astronomy, August 1991, page 34.
3. "The Grand Illusion," Astronomy, November 1992, page 44.

DEMONSTRATIONS

Drawing Ellipses

Ellipses are easily drawn as a classroom demonstration. For a piece of posterboard about 2'x3', tape it to a larger piece of thick cardboard, and use two stick pins for the two focal points. Cut a piece of string about 2' long, and tie the ends together. Start by placing both pins together at the same point, and with a colored ball point pin stretching the string tight, draw a circle with a radius of about a foot on the poster board. Now move the two pins about an inch on either side of the original center point, and with a different color pen draw a more elongated ellipse; the interfocal distance (between the pins) is now 2". Continue moving the pins farther apart, and note the new ellipses are becoming still more elongated.

. In illustrating the basic properties of the ellipse, the long axis is called the MAJOR axis, and the short axis perpendicular to it the MINOR axis. The shape of the ellipse is given by its eccentricity, or the ratio between the interfocal distance and the major axis. For a circle, as the interfocal distance is zero, so to is its eccentricity (easy to remember, since a circle looks like a zero). An ellipse so elongated that it approached a straight line would have the interfocal distance almost equal to the major axis, so its eccentricity must approach one; this to can be remembered by noting a straight line looks like the number 1. Most planet orbits have eccentricities less than 0.1, so at first glance their orbits appear as circles; only Mercury and Pluto, at opposite ends of the solar system, have eccentricity close to .25 and notably elliptical orbits. But some asteroids and most comets have much higher eccentricities.

How can we determine the eccentricity if we only know the perihelion and aphelion distances of the orbit? Note that the ellipses is symmetrical about its minor axis. The perihelion distance is as far inside one side of the orbit as the other focal point is from the other edge of the orbit. This means that the interfocal distance is just the difference between the aphelion and perihelion distance, and of course the major axis is the sum of both perihelion and aphelion, or the eccentricity is the ratio of the difference over the sum of the two values. For Mercury, for instance, (.5 - .3) / (.5 + .3) = .2/.8, or 1/4 (eccentricity = .25).

Using Skymap to compare motions of the planets

Let's compare the speeds and sizes of the retrograde loops for Mars and Saturn. Kepler tells us to expect that Mars will move faster, and inscribe a much larger retrograde loop than more distant Saturn will show. In 1997, use the track mode for Skymap (point mouse to Mars, click on right mouse button) and print a chart that will show the motion of Mars from January through July. Do the same for Saturn from June through December, 1997. Print the two charts to cover the same area of sky, and it is quite striking that Mars covers a lot more ground than does Saturn.

ANSWERS TO STUDY EXERCISES

1. The motion of the equant in the Ptolemaic model violated the principle of uniform circular motion, and so the Ptolemaic model was "...not sufficiently pleasing to the mind." Also, Copernicus was not satisfied with the accuracy of predictions made using the Ptolemaic model. (This was NOT to saw his would do any better!) Be sure that the student do not confuse the equant with, say, the epicycle. And note that Copernicus did NOT object to the epicycle as a way to explain retrograde motion--it too was circular.

2. Retrograde motion occurs when a faster moving planet overtakes a slower moving one (see the dance of the epicycles in Chapter 2 demonstrations). When Mercury and Venus overtake us near inferior conjunction, they retrograde back toward the sun as they "lap" the slower moving earth. Likewise the faster moving earth laps slower moving superior planets at opposition, so Mars, Jupiter, and Saturn appear to retrograde as the earth passes between them and the sun.

3. Jupiter would retrograde at opposition, but the earth would retrograde near inferior conjunction, when the earth passed between you and the sun. Note that at the same time an earth observer would see another planet retrograding, an observer on that planet would observe the earth in retrograde.

4. The synodic period is the time between successive oppositions or conjunctions of the same kind (the planet and earth return to the same symmetry with respect to the sun). The sidereal period is the time for planet to orbit the sun once, returning to the space position among the stars AS SEEN FROM THE SUN. Copernicus found the algebraic relationship noted in enrichment focus 3.1; for instance, if an asteroid came to opposition on May 1, 1992, then next opposition occurred on September 1, 1993, then its synodic period would be 1 year and four months, or 4/3 of a year. Then 1/sidereal = 1 - 3/4 = 1/4; the asteroid's sidereal period is four years.

5. Several answers, but one notable one was that Copernicus used it to lay out the correct scale of the solar system for the first time.

6. Copernicus considered planetary motion to be natural rotational motion of the celestial spheres, so no force was needed.

7. Essential differences were:

a. Kepler discarded the requirement for circular motion--used ellipses instead.

b. Kepler considered a force (like magnetism) determined the planetary motions.

c. Kepler's second law noted that the speed of the planet varied as its distance from the sun changed, speeding up at perihelion, slowing down at aphelion.

d. Kepler's model established general mathematical relations which ultimately can be applied to all orbiting bodies in the universe.

8. Variable answers, but Kepler's model was much more accurate than either Copernicus' or Ptolemy's. Kepler discarded the artificial geometrical symmetry, and envisioned physical forces holding bodies in elliptical orbits. Ultimately, his mathematical statement of the three laws of planetary motion would lead Newton to the law of gravity.

9. When it is farthest from the sun (at aphelion).

10. He based celestial motions on PHYSICAL causes, specifically a force linking sun and planets. His magnetism would fall aside, but it was a start.

11. When the greatest elongation occurs near Mercury's perihelion, Mercury can get only about 18 degrees from the sun. But a few synodic periods later, Mercury's greatest elongation will occur near Mercury's aphelion, and it will be about 28 degrees away from the sun.

12. Yes, asteroids obey Kepler's laws. too. Note the example in the answer to #4 above.

ANSWERS TO PROBLEMS & ACTIVITIES

1. Sin 23 deg = X / 1 AU, or X = .39 AU

2. same result as in #1

3. With Kepler's third law, $a^3 = (.39)^3 = .0593 = p^2$, thus p = .243 years or about 89 days.

4. 116/365.25 = .32 years for Mercury synodic period. So 1/sidereal = 1/.32 + 1 = 4.15, or Mercury's sidereal period is .241 years, or 88 days, in very good agreement.

5. From focus 3.3, $R_{perihelion}$ = a(1-e) = 1.52x(1 -.09) = 1.52x(.91) = 1.38 AU
$R_{aphelion}$ = a(1+e) = 1.52x(1+.09) = 1.66 AU; thus at the best oppositions, such as in 1986, Mars can get as close as .38 AU, while the least favorable oppositions (like 1995 and 1997) find Mars as much as .66 AU distant from earth. Practically, at the closest oppositions, Mars outshines Jupiter for several weeks; it will not do so in 1997 or 1999, but will by the much closer opposition in 2003.

22

6. Jupiter's sidereal period is found via Kepler's third law; $a^3 = (5.2)^3 = 140.6 = p^2$; thus the sidereal period of Jupiter is 11.8 years.

7. For a Mars-based observer, the greatest elongation angle would be calculated as $\sin X = 1$ AU/1.52 AU or .65, so the maximum elongation of earth would be 41 degrees from the sun.

8. As seen from Saturn:
a. the earth would come into inferior conjunction with the sun every 378 days, the same as the time interval for us to see Saturn come to successive oppositions.
b. the earth would reach a maximum elongation of $\sin x = 1/9.54 = .104$, or about 6 degrees.
c. Saturn's synodic period is $378/365.25 = 1.035$ years, so $1/\text{sidereal} = 1 - 1/1.035 = .0337$, so Saturn's sidereal period is 29.6 years.

9. Mars has an orbit $1.52 \times 2 \ (3.1416) = 9.55$ AU in circumference, which it covers in 2.2 years, so Mars travels at 4.34 AU per year. Jupiter has a circumference of 32.6 AU, but it takes Jupiter 11.8 years to complete its trip, at a speed of 2.76 AU per year. Finally, Saturn's orbital circumference is 59.94 AU, but it takes Saturn 29.6 years to orbit, at 2.02 AU per year. Obviously the farther out a planet is, the slower its orbital motion around the sun.

CHAPTER 4

The Clockwork Universe

CHAPTER OUTLINE

Central Question: How did Newton's laws of motion and gravitation furnish a unified, physical model of the cosmos?

4.1   Galileo: Advocate of the Heliocentric Model
     A. The magical telescope
     B. The Starry Messenger
     C. Galileo discoveries and the Copernican model
     D. The Crime of Galileo

4.2   Galileo and a New Physics of Motion
     A. Acceleration, velocity, and speed
     B. Natural motion revisited
     C. Forced motion: gravity
     D. Galileo's Cosmology

4.3   Newton: A Physical Model of the Cosmos
     A. The prodigious young Newton
     B. The magnificent Principia
     C. Forces and motions
     D. Newton's laws of motion

4.4   Newton and gravitation
     A. Centripetal acceleration
     B. Newton's law of gravitation

4.5   Cosmic Consequences of Universal Laws
     A. The earth's rotation
     B. The earth's revolution and the sun's mass
     C. Gravity and orbits
     D. Orbits and escape speed
     E. Newton's cosmology

Enrichment Focus 4.1: Speed, Velocity, and Acceleration
Enrichment Focus 4.2: Newton, the Apple, and the Moon
Enrichment Focus 4.3: The Mass of the Sun
Enrichment Focus 4.4: Escape Speed

CHAPTER OVERVIEW

Teacher's Notes

Although the mathematical details are beyond many (if not most) students in introductory astronomy courses, make the point that the full set of Kepler's three laws requires the gravitational force to follow an inverse-square law. Newton's gravitational law (with Kepler's laws) allows the determination of the sum of masses in an orbiting system, or the central mass if it is much greater than the orbiting mass. One question which arises is, how can we determine the mass of the moon? Methods used in the past involve measuring the displacement of the center of mass of the earth-moon system from the earth's center, analysis of tidal effects, or assumptions about the moon's composition. This is discussed more fully in chapter 9.

To see the dependence of Kepler's Third Law on the central mass, students may plot values R^3 versus T^2 (or R versus T on a log-log plot, as in the text) for different systems, say, Jupiter's satellites and man-made satellite orbiting the earth. The graph will show all the points for a given system falling along a straight line, but different lines arise for different systems.

One term in common usage that sometimes causes confusion or misconceptions when dealing with Newtonian gravitation is "weightlessness' as applied to an orbiting objects, for instance, an astronaut in orbit about the earth. In Newton's treatment of gravitation, weightlessness does NOT mean that no force is acting on the astronaut; in fact, essentially, the ONLY force acting on a "weightless" astronaut is his weight, or the attractive gravitational force due to the earth. Perhaps free fall describes this motion or state better than weightlessness. This apparent contradiction in terms will lead nicely to the material in chapter 7.

Along the same lines, the description of an object acted upon by the gravitational force alone as being in free-fall, when applied to the moon, means that the moon is constantly falling toward the earth's center. It may be interesting to tell the students that your observed the moon to be falling toward the earth, and have them discuss the validity of that statement in terms of Newton's theory of gravitation.

Be warned that students may seem to catch the idea of an inverse-square law when in fact they have simply learned to square numbers. Having them graph an inverse-square law for some sample points is well worth the time! They show all understand this graphical relationship.

4.1 *Galileo: Advocate of the Heliocentric Model*

In his discussion of Venus' phases, Zeilik should probably contrast the entire set of phases seen by Galileo with the prediction of Ptolemy's epicycles. Ptolemy had Venus' (and Mercury') epicycles always between us and the sun, so according to Ptolemy, Venus should always have appeared as a crescent through a telescope of 30x of higher. The fact that Venus went all the way around the sun with its own independent orbit, in no way governed by the earth's position, was perhaps the most telling refutation of the geocentric model and Ptolemy's epicycles.

4.2   Galileo and a New Physics of Motion

With Galileo's physics, we see a stress on experiment and careful observation. Galileo's work on inertia essentially leads to Newton's first law of motion. Galileo's great mind and combative spirit are well illustrated by his house arrest. With failing eyesight and in hot water with the church over his telescopic discoveries and books reporting them, he turned instead to practically invent modern physics. How many of us, faced with such roadblocks, would have been able to triumph in the birth of both modern astronomy and physics?

4.3   Newton: A Physical Model of the Cosmos

The distinction between mass and weight can be made in discussing the second law; the mass depends on how much substance the object has, while its weight is also determined by the surrounding gravitational field. On the earth, the object's weight is $W = m \times g$; the same object would have the same mass on the moon, but with the moon's gravity only a sixth as strong as ours, only weigh a sixth as much as on earth.

For many students, the third law is hard to comprehend. It is saying that the gravitational force of the earth on the sun is identical to the force the sun exerts on us. Why then is it the earth which orbits the sun? The first law relates the mass to the object's inertia, or resistance to change of motion. As the sun is about 333,000 times as massive as the earth, the second law tells us that with equal but opposite forces, it is the earth that is accelerated 333,000 times more than the sun! That is why we call it the solar system; the sun's great mass and gravity is basically calling the shots.

4.4   Newton and Gravitation

Most students do not seem to have a problem with the masses affecting the gravity of bodies; if we were on a planet which was the size of earth, but with twice its mass, just about everyone knows we would weigh twice as much. This was at least suspect well before Newton's time. Newton's great break-through was the inverse SQUARE relationship; most suspected that gravity declined as bodies drew apart, but Newton showed the exact relationship. Note that with the distance factor being squared, a slight change in distance can become a substantial change in gravitational force. Point to Zeilik's example of the moon's acceleration toward earth.

4.5   Cosmic Consequences of Universal Laws

In using Newton's revision of Kepler's third law, note that it would be appropriate to approximate the masses with just the mass of the great body if the satellite is far less massive than its primary. Zeilik is certainly right in neglecting our mass compared to the sun's in his example, and you could likewise use any satellite to weigh its planet safely except for Pluto, for Charon is half as large as its planet and probably at least a tenth as massive.

But as the two bodies become more similar in mass, you must first just calculate the total mass of the system, then observe their interaction to find the common center of mass, and divide the total mass up between the two bodies of this cosmic see-saw. For instance, many asteroids may be binary; Toutatis in particular may be two equal bodies almost touching in close orbit around a common center of gravity; see Astronomy for January 1995, page 30 for more some interesting cases.

In a binary star system, the more massive star would be revolving slower and be located closer to the barycenter (common center of gravity); if the total mass was three suns, and the larger star was twice as close to the barycenter, it must weigh two solar masses, and its companion is about as massive as our star.

Escape velocity is about 11 km/sec, or about 25,000 miles per hour. Note it is not much faster than the LEO velocity of 18,000 mph for Space Shuttle. If we want to compare the earth's escape velocity with that of any other body, we need only find the comparative ratios of masses and SQUARE the ratio of their radii. For instance, Mars has a mass of only 1/10th the earth, and is half as large. Thus the gravitational force on Mar's surface is going to be $.1/(.5)^2$, or about 4/10ths as large as ours. Applying this to our Moon and most smaller bodies, we find their surface gravities too tiny to hold any atmosphere. Much more massive jovian planets do not have surface gravities much greater than earth's, but their distance from the sun keeps them cool enough to hold onto light hydrogen and helium, explain why they are so much larger.

ADDITIONAL RESOURCES

Transparencies 4.5, 4.12, 4.13, 4.14ab, 4.16abc, 4.17, 4.18, 4.19, 4.20, and 4.23 are available to assist with your class presentation of the material.

SUPPLEMENTARY ARTICLES

1. "Building a Copyscope for under $25," Astronomy, May 1986, page 74.
2. "Fun Telescopes for under $10," Astronomy, May 1987, page 46.
3. "Looking for Planet X via Gravity," Astronomy, August 1988, page 30.
4. "The History of the Grand Tour," Astronomy, September 1989, page 44.
5. "Is the Solar System in Chaos?," Astronomy, May 1990, page 34.
6. "Your first date with a 2.4" Telescope," Astronomy, January 1993, page 90.
7. "Cheap Shots," Astronomy, August 1993, page 30.
8. "Double Asteroids mean Double Trouble?," January 1995, page 30.
9. "The Remarkable Odyssey of Jane Luu," February 1996, page 46.
10. "The Discovery of Neptune," September 1996, page 42

ADDITIONAL DISCUSSION TOPICS

Geosynchronous Orbit

One of my favorite discussions which many students can relate to is how Arthur C. Clarke and Ted Turner destroyed communism. In 1946, Clarke introduced the concept of putting communication satellites into geosynchronous orbit. Begin by asking the students about what the orbital period of a satellite in low earth orbit (LEO) such as Space Shuttle is. Some may know it is about 90 minutes. Then ask them what would happen if it were placed in a higher orbit.

Most students should see that farther from the earth's center, the gravity would be decreased, so the corresponding centripetal force decreased as well; the satellite would not need to orbit as fast, so it would take longer to travel the increased distance at a slower speed.

Ask them what would happen to the gravity if you chose an orbit of about 4 earth radii; they should be able to calculate that the gravity would drop to 1/16th g; thus as Clarke found, at this distance the orbital period would be 16x greater, or 24 hours. Clarke realized that such a satellite would apparently hover over a single point above the earth's equator in such a "geosynchronous" orbit, revolving around the earth at exactly the rate we rotated beneath the satellite.

Thus our communications dishes can be pointed to such hovering satellites, and TV, telephone, and other data constantly exchanged between ground stations at all times, the basis for HBO, CNN, and the entire cable industry, and the communications revolution that has made the human race a much closer and better informed family than anyone could have dreamed at the end of World War II.

The Grand Tour

We have lived in a most historic time. Thanks to a special arrangement of planets in the late 1970's, our Voyager I and II space probes were able to fly past all four of the outer jovian planets, with the fantastic results the students will see later in the book.

The physics of how this was possible is an excellent illustration of Newton's laws at work. As the Voyagers neared Jupiter, they both passed BEHIND it, causing the giant planet to slow down a little bit (about a foot per trillion years, according to NASA); the equal but opposite force of Jupiter on the tiny Voyagers caused both of them to be accelerated a LOT, with a considerable gain in speed and direction---on to Saturn. While Voyager I was "sacrificed" in order to study Titan and Saturn's rings from above the ecliptic plane, Voyager II stayed in the ecliptic and was redirected by Saturn on to Uranus, and by Uranus on to Neptune. Unfortunately the alignment did not allow inclusion of Pluto, but a mission to Pluto, using Jupiter for a gravity assist, is possible early in the 21st century.

There is such a thing as a free lunch; you must just be at the right place at the right time to cash in. Voyager II is the extreme case of a gravitational free loader. Had it relied on its own Titan booster and Centaur upper stage to get it on the shortest, fastest course to Neptune back in 1977, it would have arrive at the most distant jovian in 2010. By taking what would seem to be a far more circuitous route, past the other three jovians, it arrived 21 years ahead of schedule! This is because each planetary encounter gave Voyager more gravitational energy and sped it up immensely faster than any human engineering could have.

It is worth noting here that Voyager is NOT a satellite, as most students would think. Remind the students that satellites are traveling at orbital, but not escape velocities--they are in orbits around their planets. Objects that are not so bound are PROBES, making one time passes such as Voyagers did with the outer planets. They are both traveling faster than the escape velocity of even the Sun, so they are now starships, on infinite voyages, trapped only the gravity of the Milky Way itself.

In July 1993, Voyagers sent back an additional benefit from their star-bound trajectories. The magnetometers aboard the probes reported the probes, now out at about 40 AU, were approaching the heliopause, the outer edge of the sun's magnetic field. By 2010 they will be starships in space, floating out in the random currents of the interstellar medium.

Predicting and Observing Bright Earth Satellites with the Internet

One of the most striking examples of gravity in action is the regular return of bright satellites in low earth orbit. The Russian space station Mir (with American crew aboard), our Space Shuttles, and the Hubble Space Telescope are all larger than a tour bus, and easily seen with the naked eye, if you can find out when to be watching and from what direction. All appear to be moving west to east; it is much easier to launch a satellite in orbit if your rocket is launched eastward, taking advantage of the earth's over 1,000 mph rotation. You will, incidentally, see some bright satellites in polar orbit as well, but it is much harder to get information on these. They are secret military spy missions, and you will not find information on their constantly changing orbits posted. Even the times of appearance of the Mir, Shuttles, and Hubble are subject to change--don't be surprise if they are several minutes early or late as ground controllers often must adjust their orbits.

The University of California at Berkeley runs one of my favorite hit sites with their "Satpasses" home page, with listings for many cities around the world for the brightest satellites. In many cases, there will be no observable passes for the coming week, or all the passes will be at night (NNNNN) or during the day (DDDDD), but for local times of 0500-0700 (early morning) and 1900-2100 (evening twilight), you may find passes listed as NVVVV, with the V indicating it should be visible from the location asked for. To access the satpasses homepage, you can use http://ssl.berkeley.edu/isi_www/satpasses.html, or just key in "satpasses" on your web browser. Try to let your students know days in advance when to watch for the passes.

These are especially neat when the Shuttle Atlantis is going to chase and join with the Russian Mir. When so linked, they form the largest manmade object ever to orbit the earth. I remember July 4, 1995 when about 8:20 PM CDT, they paraded past the thousands assembled in Pensacola to watch the fireworks to follow soon after. They had already separated, and were about a minute apart (or about 25 degrees apart in the sky) but still in basically the same orbit, and the stately parade of the two brightest objects in the evening sky was something which made that Independence Day very memorable to many folks.

The Hunt for Neptune

It has been said the discovery of Neptune was the greatest triumph of Newtonian physics. Certainly it is a great yarn of international cooperation and controversy. For fine stories of geniuses at work, uppity servants, overworked observers, and missed opportunities, read the accounts of the search in the September 1996 issues (marking the 150th anniversary of Neptune's discovery) of both Sky and Telescope (page 42) and Astronomy (page 42).

The physics involved is most interesting. From Herschel's accidental discovery of Uranus in 1781 until 1822, the seventh planet ran faster than Newton would have predicted, based on the gravities of the Sun, Jupiter, and Saturn. Just as mysteriously, after 1822 it slowed down. Both John Couch Adams of England in 1845, and a year later Urban Levierrer of France independently predicted where the next massive planet must lie in Aquarius to created the observed perturbations of Uranus. The English search, based on Adam's unpublished letter to Airy, was a comedy of errors, but Levierrer's published predictions allowed the German observer's Galle and D'Arrest to locate the outer jovian within thirty minutes of starting their search. All students of the history of science will find many lessons in this great story.

DEMONSTRATION & OBSERVATIONS

Galileo's Telescope Updated

Edmund Scientific sells a simple, nonachromatic refractor kit which approximates Galileo's first telescope; look for their adds in astronomy and science magazines. At 8x it is lower in magnification than some of Galileo's scopes, but much easier to use.

It will still show the students Jupiter's moons, Venus' phases, sunspots (via projection, please!), and many deep sky objects. It can be assembled by the students as a class project, and is available at classroom discounts. Still, many students will note how much better image quality even a pair of achromatic binoculars gives than the instruments Galileo used--it is amazing just how much he was able to observe.

Building a Gravity Well

Do you have one of the coin collectors that look like a sloping funnel, where the coin spirals down to the model black hole? If so, use it in class, noting the centripetal force as holding the revolving coin in orbit and being centrally directed, as viewed from the top. A small marble would probably be an even better choice for your model planet in this demonstration; call on some students who rarely get involved in discussion to assist with the demo, dropping the marbles into the gravity well from a variety of directions and speeds.

The Dance of the Jovian Moons

Observing the revolution of Jupiter's four large moons around the giant planet is a great way to let the students note the cosmic implications of Newton's laws. Innermost Io will switch from one side of Jupiter to the other night after night, with a revolution period of less than two days. By contrast, outermost Callisto takes over two weeks to complete its much longer circuit of the giant world. If night observing is out, make copies of the current motions of Jupiter's moons for the entire month from the latest copy of Ottewell's Astronomical Calendar, or the current Astronomy or Sky and Telescope, and encourage students to get out their binoculars or small telescopes and observe this planetary dance for themselves.

The program "Dance of the Planets" allows graphic depiction of the jovian satellite system, with the ability to zoom in on Jupiter (128X allows good views of all four large Galilean moons) and run time forward. It is especially fun to use the program to predict events that may be observed later that same evening. Allow students to use it for extra credit, then observe their predictions happen.

ANSWERS TO STUDY EXERCISES

1. Several good answers here, such as:
a. Jupiter's four moons orbit Jupiter, not the earth, as required in a geocentric model.
b. Venus' phases prove it orbit the sun, again refuting the geocentric model.
c. the sunspots move across the sun's face as it rotates, proving a smaller body like earth could also be rotating to produce diurnal motion.
d. the craters and mare of the moon showed it an imperfect body.

2. Galileo ignored Kepler's mathematical description of celestial motions (elliptical and nonuniform) and Kepler's three laws of planetary motion, and instead concentrated on a description of terrestrial motion to find physical ideas that would support the Copernican system.

3. Galileo's concept of inertia reversed the Aristotelian view of forced and natural motion. So in Galileo's view an object moving on a level surface (without friction) would continue to move in a straight line indefinitely with no force applied to it.

4. Both balls fall with the same acceleration (neglecting frictional effects) so that they fall through the same distance in equal times, and so they would both hit the ground at the same time if dropped from the same height at the same time.

5. You and the object both move away from the point in space at which you pushed the object. You move in the opposite direction and with a final speed equal to the object's final speed (Newton's third law), since you both have the same mass and experienced the same force, and so the same acceleration.

6. Newton's model included a force that governed the motion of orbiting bodies, whereas Copernicus' model did not require forces from planetary motion. Also, Newton's cosmic system was infinite in extent, rather than finite.

7. The traditional objection to earth's rotation was that objects thrown upward would land behind their starting point. Newton responded that the forward motion of the object (in the direction of our rotation) continues when the object is aloft because of its inertia, so it drops back at the starting point. In terms of our revolution, it is the sun's gravity that keeps the earth on an elliptical path.

8. Kepler and Newton agreed on elliptical orbits for planets. Kepler considered a magnetic force holding the planets in orbit, while Newton found gravity did the trick.

9. Many examples are good, such as:
First law: Voyager's traveling in a straight line in the interstellar medium
Second law: If a Volkswagen and Cadillac collide, the bug ends up in the ditch
Third law: Jupiter was slowed down a tiny bit in order for Voyager to be accelerated

10. The ball lands AHEAD of the target, for its inertia keeps its forward motion after its release from your hand. In fact, the ball will keep up with you and land at your feet. Many students will believe that the ball falls BEHIND the target. Do this as a class demo!

11. The object rises, moving more and more slowly, until it hits a maximum height at zero speed. It then falls downward, accelerating until it hits the ground as the same speed that it was shot off (if the air friction is small).

12. The centripetal accelerations are the same. Students will confuse this concept with that of force; the gravitational forces on the moon and the satellite are different.

13. They are the same (Newton's third law)! Some students will believe that, since Mars has less mass then the earth, it would exert a lesser gravitational force.

14. As we go out from the sun, the square of the planet's orbital period in years and proportional to the cube of its average distance from the sun in astronomical units.

ANSWERS TO PROBLEMS & ACTIVITIES

1. Working out the forces in relative terms, we use $F = G.M./R^2$, and let both G and M (the sun's mass) be one for simplicity, thus for earth, the force would be one unit force.
Mercury: $(.055) / (.39)^2 = .055/.152 = 0.36$
Venus: $(.82) / (.72)^2 = 1.6$
Mars: $(.11) / (1.52)^2 = .048$
Jupiter: $318 / (5.2)^2 = 11.8$
Saturn: $95 / (9.5)^2 = 1.1$
Uranus: $14 / (19.2)^2 = .038$
Neptune: $17 / (30)^2 = .019$
Pluto: $.0002 / (39.4)^2 = 1.3 \times 10^{-6}$

2. Letting the earth's acceleration again be one, and as $F = ma$, divide the force found in # 1 above by each planets mass, so we get:
Mercury: $.036 / .055 = 6.5$
Venus: $1.6 / .82 = 2.0$
Mars: $.048 / .11 = .44$
Jupiter: $11.8 / 318 = .037$
Saturn: $1.1 / 95 = .012$
Uranus: $.038 / 14 = .0027$
Neptune: $.019 / 17 = .0011$
Pluto: $1.6 \times 10^{-6} / .0002 = 6.5 \times 10^{-4}$

3. $V_{escape} = (2GM/R)^{.5} = (2 \times 6.67 \times 10^{-11} \times 7 \times 10^{22}/1.7 \times 10^6)^{.5} = 2.4$ km/sec

4. The initial acceleration is $a = F/m = 10N/1 \text{ kg} = 10 \text{ m/s}^2$
With five kilograms, we get $10N/5 = 2 \text{ m/s}^2$
With ten kilograms, we get $10N/10 = 1 \text{ m/s}^2$

5. For simplicity, use two football players (100 kilogram = 220 pounds); then
$F = G.M./R^2 = (6.67 \times 10^{-11} \times 100 \times 100) /(10m)^2 = 6.67 \times 10^{-9}$ N

6. Centripetal acceleration $= v^2/R$, so the velocity is orbital circumference / 90 minutes
$v = (3.1416 \times 1.2756 \times 10^7 \text{ m}) / 90 \times 60 = 7,421$ m/s; $a = (7,421)^2 / 6.378 \times 10^3 = 8.635 \text{ m/s}^2$, similar to the 9.8 m/s^2 for the earth's gravitational acceleration.

7. $P^2 = (90 \times 60)^2 = 2.92 \times 10^7 = 4 (3.1416)^2 (6.5 \times 10^6 m)^3 / 6.67 \times 10^{-11} \times (M_{earth} + M_{satellite})$
Since the mass of the satellite is negligible, this gives us the mass of the earth as:
$M_{earth} = 39.5 \times 2.75 \times 10^{20} / 6.67 \times 10^{-11} \times 2.92 \times 10^7 = 5.6 \times 10^{24}$ kilograms.

8. At opposition, Jupiter is about 4.2 AU distant, or 630 million km away. To find the separation angle, $\sin x = 1.1 \times 10^6 / 6.3 \times 10^8 = .001746$, so the maximum separation is about 6 minutes of arc, well within the range of human visual resolution.

CHAPTER 5

The Birth of Astrophysics

CHAPTER OUTLINE

Central Question: How do atoms produce light, and what does light reveal about the sun and stars?

5.1   Sunlight and Spectroscopy
      A. Atoms and matter
      B. A model of the atom
      C. Simple spectroscopy

5.2   Analyzing Sunlight
      A. Kirchoff's rules
      B. The conservation of energy
      C. Kinetic energy
      D. potential energy

5.3   Spectra and atoms
      A. Light and electromagnetic radiation
      B. Waves
      C. The electromagnetic spectrum
      D. Atoms, light, and radiation
      E. Solving the puzzle of atomic spectra
      F. Energy levels
      G. Other atoms

5.4   Spectra from Atoms and Molecules
      A. Absorption-line spectra
      B. Spectra of molecules

Enrichment Focus 5.1: Kinetic energy

Enrichment Focus 5.2: Energy and Light

CHAPTER OVERVIEW
Teacher's Notes

Concepts of wavelength and color will be familiar to most students who have had a physics or chemistry course at any level (few of them may have had neither); however, they may not have observed spectra. Spectra are difficult to visualize if not observed directly. It may be hard to relate photographs of spectra, especially black and white prints, to the pattern of colored lines and band one sees in a spectroscope. (See the color plate of spectra in the text; draw your students' attention to it.)

If at all possible, have students observe spectra using the spectroscope provided with the eighth edition of this text. Focus 5.1 contains instructions for using this simple spectroscope. Geissler tubes containing different gases may be easily observed through a diffraction grating film. An interesting exercise is to have students determine the chemical element that is inside fluorescent light by comparing a fluorescent light spectrum (continuous with bright lines) with the spectra of different Geissler tubes, one of which contains mercury vapor.

Absorption spectra are more difficult to observe directly; calcium and sodium lines can be seen in sunlight using a simple diffraction grating spectroscope. Caution students not to point a spectroscope directly at the sun but to view sunlight reflected from a sheet of white paper or a white cloud). If it is not possible for students to observe spectra themselves, try to demonstrate spectra. The sun's spectrum can be projected onto a screen. That the problem of sorting out the lines in spectra is not a simple one can be shown by using the solar spectrum as an example. There are at least twenty thousand cataloged spectral lines in the solar spectrum. Slides of the solar spectrum can be obtained from several sources.

This chapter may be skipped without breaking the development of ideas about motion and gravitation that are expanded in chapter 7. It MUST be done before chapter 13. However, to handle chapter 7 completely requires at least an acquaintance with dark-line spectra of stars and the fact that the dark lines occur at specific wavelengths (in the lab) for each element.

NOTE: Some student are very confused that astronomers call a continuous spectrum with a dark line an "absorption spectrum." They tend to think of it and call it a "continuous spectrum," which results in confusion about the basic types of spectra.

The most difficult concept in the chapter is that of energy levels of atoms. Emphasize this point with the figures. The analogy of the stairs works very well and should be repeated.

5.1 Sunlight and Spectroscopy

No one has ever touched a star. Everything we know about the nature of stars must come from the electromagnetic radiation that travels through great distances in the near vacuum of space to reach us. For stars, at temperatures of thousands of degrees kelvin, a principal form of that radiation is the one our eyes are tuned in on, visible light. With the science of spectroscopy, we have learned a great deal about the stars.

The fundamental difference between stars and planets is temperature. The 6,000 K photosphere of the Sun is about 20X hotter than the approximately 300 K temperature of the classroom. It is thus indeed a very good thing we are not touching a star!

Looking ahead to the discussions of the densities of white dwarfs and neutron stars, it is worth noting in this review of basic chemistry that the atom is chiefly empty space. Note the electrons are 2000x less massive than either protons or neutrons. The nucleus is tiny compared to the electron orbital, but contains almost all the mass of the atom. A good analogy is the solar system, with the sun containing practically all the mass, the planet and comet orbit occupying most of its volume; like the solar system, the atom is chiefly empty.

It may be old hat, but find out if anyone in your class hasn't been introduced to ROY G. BIV in getting down the colors of the continuum. Be sure your students grasp that this continuum is the norm; bright or emission lines stand out brighter than the continuum, while dark or absorption lines appear so because cooler atoms have absorbed some of the light.

5.2 Analyzing Sunlight

Stress in Kirchoff's first rule that density is critical to observing the continuum. It is created by the collisions of many atoms in a dense medium; these collisions happen at a wide variety of energies; violent collisions produce photons of violet light, while low energy collisions (fender benders on an atomic scale) make tail-light red photons. If the atoms become so widely spaced that such "thermal" collisions become rare, then the energy changes we will observe occur not between atoms, but within their electron orbital. Electrons losing energy give off emission lines (Kirchoff's second rule) while those absorbing energy are revealed by a dark line spectra (Kirchoff's third rule).

5.3 Spectra and atoms

In discussing the electromagnetic spectrum, note the ranges of AM and FM radio frequencies; AM is measured in kilohertz, or thousands of waves/second, while FM is given in megahertz, with millions of waves/sec. Have the students calculate the carrier wavelength of an FM station with a frequency of exactly 100 MHZ--there is probably one close to that in your area. As wavelength = c/f, then 300,000,000 meters/sec / 100,000,000 waves/sec yields just 3 meters, or about 10 feet for this carrier wave.

Your students are probably not familiar with angstroms. They were developed in chemistry for measuring atomic diameters; the neutral hydrogen atom at ground state is about an angstrom across. A sheet of paper .1 mm thick would be a million angstroms deep.

In discussing the Bohr model of the atom, your students may wonder why the ground state of hydrogen is not involved. In fact, the transitions to this lowest level do occur in greater profusion than to the Balmer series we see visually. But as Figure 9.8 shows, the energy change down to ground state is far greater than in the Balmer transitions. These Lyman spectral lines all occur in the ultraviolet, beyond our visual perception. Likewise the transitions from the third level upward are even less energetic; the Paschen transitions must be detected in the infrared. Again, what we see is the proverbial tip of the iceberg.

In the Balmer series, the 3-2 transition produces the bright red hydrogen alpha line in the spectrum. As there is a lot more hydrogen than any other element, this is the dominant color of the great nebulae of ionized hydrogen (H-II regions) like M-42, the Orion Nebula of winter, or M-8, the Lagoon Nebula of Sagittarius in the summer Milky Way. It also is very obvious at solar eclipses as the red color of the chromosphere and prominences.

Your students may note that the 4-2 transition (hydrogen beta) carries more energy (a line in the blue-green) than does the red hydrogen alpha transition. Why then is red the dominant color of the cosmos? Statistically there are many more electrons in the third level falling down to the second, producing the hydrogen alpha line, then there are excited up to the fourth level, making hydrogen beta radiation as they drop in energy.

In terms of physics, relate excitation to electrons moving to higher orbits, while ionization represents escape velocity for the electron, leaving its nucleus behind. But unlike satellite orbits, the electron orbital are quantized, and only a few are stable.

5.4 Spectra from Atoms and Molecules

To understand why the various elements have different spectral lines, remind your students that the spectral lines are produced in the electron orbital of the atoms, the same places where chemical reactions occur. Each element is defined by its chemical reactions because it has a different electron orbital configuration than any other element possesses. Thus each element has a unique set of electron transitions which produce its own set of spectral lines, in essence, its spectral fingerprint.

ADDITIONAL RESOURCES

The slide of the spectral image beginning this chapter and transparencies 5.2, 5.4a-g, 5.5abc, 5.6, 5.7, 5.9abc, 5.10abc, 5.14, and 5.17 will supplement class presentations.

The ninth episode of Cosmos, "The Lives of the Stars", is highly recommended for Unit II; it could be shown here to introduce the unit, or shown later in the chapters on stellar evolution. But use it--there is nothing dated about any of its material, and Sagan does an outstanding job of relating the stars and their lives to chemistry and life itself. In fact, this video would not be a bad introduction to a chemistry course.

SUPPLEMENTARY ARTICLES

1. "Photoelectric Photometry--to better measure starlight," Astronomy, June 1986, page 74.
2. "Reading the Colors of the Stars," Astronomy, April 1989, page 36.
3. "The Hottest Stars in the Universe," Astronomy, February 1990, page 22.
4. "A Journey with Light," Astronomy, March 1990, page 30.
5. "The Coolest Stars," Astronomy, May 1990, page 20.
6. "Magnificent Orion," Astronomy, November 1990, page 78.
7. "Seeing the universe in the infrared," Astronomy, June 1991, page 50.
8. "GRO: The Violent Universe in Gamma Rays," Astronomy, July 1991, page 44.
9. "The Faintest Stars," Astronomy, August 1991, page 26.
10. "The Smallest Stars in the Universe," Astronomy, November 1991, page 50.
11. "The Brightest Stars in the Galaxy," Astronomy, May 1991, page 30.
12. "The Birth of Radio Astronomy," Astronomy, July 1992, page 46.
13. "Fixing Hubble's Vision," Astronomy, February 1993, page 42.
14. "The Next Twenty Years in Astronomy," Astronomy, August 1993, page 29.
15. "Big Scopes for the 21st Century," Astronomy, August 1993, page 48.
16. "How to Beat Light Pollution," Astronomy, September 1995, page 44.

ADDITIONAL DISCUSSION TOPICS

The Audible versus Visible Spectrum

Most of us value our sight far more than our hearing, but any blind person can testify that we vastly underestimate the value of our ears. For instance, our ears respond to a far wider range of frequencies than do our eyes. You probably have some music majors in this class of mostly nonscience majors, so ask if any can define the octave in music. Some will respond eight notes, but many one will know that as you go up an octave, you have doubled the frequency and thus inversely halved the wavelength of the sound waves. Next ask how many octaves you can hear; if they know human audible acuity, they find from about 40 Hz up to 16,000 Hz is a little more than eight octaves.

Now turn to the range of human vision. We see from violet light at 4,000 A to red light at 7,000 A. Is this an octave? Not quite, only about 3/4 of an octave, or the equivalent of just six notes (do to la on the keyboard or vocally). Yet until the 1920's everything we knew about the universe beyond earth had come to us via just those six notes.

In this century we have at last gained access to all those missing notes. In the 1920's Jansky's pioneering work at Bell labs opened up the bass section of the cosmic orchestra to our examination via radio waves. But because of the blocking of most E-M radiation by our atmosphere, other windows had to wait for Sputnik to be opened. The Copernicus Satellite Observatory in the 1960's first opened the ultraviolet (the alto section) to our hearing, while Uhura in the mid 1970's tuned in on the X-rays and gamma rays (the soprano notes). IRAS in 1982 examined the infrared universe (the tenor notes), and in 1989 COBE filled in the last gap with the microwaves (the baritone section). For astronomers, it is indeed a time of joyful rejoicing--at last we can hear the entire cosmic symphony. Each time we opened a new window of E-M radiation for examination, new and strange and wonderful objects stared back at us...quasars, pulsars, black holes, neutron stars, brown dwarfs, things for which we have not yet invented names. This, more than any other reason, is why we must change astronomy text books or at least get new editions every 2-3 years. In astronomy, new data from a variety of wavelengths pours forth in truly astronomical quantities.

Gamma Ray Burstars and Cosmic Rays

The extreme high end of the EM spectrum's frequency and energy range is full of mystery. The origin of the majority of the cosmic rays Sagan discussed in the Cosmos episode 109 now appears to be the optically invisible Supernova Remnant in Lupus, remains of a type Ia SN of 1006 AD. Even more mysterious is the apparently homogeneous distribution of gamma ray burstars, discussed in the September 1996 issue of Sky and Telescope on page 32. Everything from colliding comets to neutrons stars to decaying pulsars or black holes are being considered.

DEMONSTRATIONS AND OBSERVATIONS

Class Spectroscopy Lab

If you plan to let your class build the simple spectroscope at the end of this chapter, Edmund Scientific Company is a good source of class sized packets of inexpensive diffraction gratings. Call (609) 573-6250 if your department doesn't already have their latest catalog.

In his discussion, Zeilik does not give a good source for an emission spectrum you probably have in your classroom...your florescent lights. As the students lift their spectroscopes, they will see a pattern of bright indigo, chartreuse, and dingy yellow lines which are produced by the mercury vapor inside most florescent lights. Also, to see the sun's spectrum, you do not need to look at the sun. Sunlight is all around you, and any bright white object (like a cloud, for instance) makes a fine reflector to view the sun's spectrum. Your students may have trouble picking out the absorption lines at first; give them the hint of looking in the yellow for the sodium D line, the easiest to spot.

If you don't want the whole class to build the spectroscopes provided in the eighth edition of the text, then just pass around a good quality commercial one in class, especially if you have sunlight coming in through your windows and florescent lights overhead. As noted above, they can observe all three types of spectra in class. If it is at night, and the moon is bright, the moon serves as a good reflector of sunlight and will also yield a good solar spectrum.

Quantum Leaps

While not as adept at this as Scott Bakula in TV's "Quantum Leap," I also illustrate the concept to my students in class. The floor, a sturdy chair, and an even sturdier desk will suffice. I, an electron, spend most of my time on the floor, in ground state. But I get excited, and lead up to the seat of the chair. I absorbed energy to do this, while increasing my potential energy in the transition. With more excitation, I can even leap up to the table top (you may not trust you table, and decide to stop at the chair). But there are only three stable orbital available to me. Waving my leg in empty space, I point out there is no stable orbital there, and then jump from the table to down to the floor. In this drop, I released a photon of infrared energy (the soles of my shoes are now hotter) as an emission line. Similarly on a quantized ladder, you need to be on one rung or the other--hanging around between rungs is not recommended.

ANSWERS TO STUDY EXERCISES
1. You see an emission (bright-line) spectrum, due to the hot, thin sodium gas. The only lines visible are two bright yellow lines, the sodium D lines that are so obvious in the sun's spectrum.

2. We observe the sodium doublet in the sun's dark line spectrum at exactly the same wavelengths and energies found in the lab on earth.

3. In the cooler outer atmosphere of the sun, the thin gas atoms absorb at the characteristic wavelengths of each element in lab plasmas on earth. Almost all elements found naturally on earth have been also found in the sun's spectrum.

4. As neutral hydrogen had only one electron, ionized hydrogen has lost its only electron and so shows NO spectral lines at all. All spectral lines depend on energy transitions between orbital, so if no electrons are present, there are no transitions nor any lines.

5. Absorption lines are produced when electrons absorb energy, jumping up to a higher energy level from a lower one.

6. Radio, infrared, ultraviolet, and X-rays.

7. The moon is merely reflecting sun light, without adding any features of its own (it lacks an atmosphere to absorb any additional wavelengths).

8. If a hydrogen atom looses its electron, it is ionized, and has a net positive charge.

9. You would see two different emission line spectra superimposed.

10. The kinetic energy is greatest when the planet is moving fastest, at perihelion, and least at aphelion. The total kinetic energy is given by $\frac{1}{2} m v^2$, where v is the average velocity.

11. All forms of light and other electromagnetic radiation travel at 300,000 km/sec or c.

12. The longest wavelengths of EM radiation lie in the radio portion of the spectrum.

13. When a photon is absorbed, an electron jumps up from a lower to higher energy level, and a dark line is seen in the spectrum due to that photon not reaching us.

ANSWERS TO PROBLEMS & ACTIVITIES

1. The H and K lines of calcium are about 3,950 Angstroms, while the sodium D line is about 5890 Angstroms. The calcium lines are thus shorter in wavelength and so higher in energy; the actual ratio is 1.49x more energetic for calcium.

2. wavelength x frequency = velocity (c for all electromagnetic waves) = 300,000 km/sec.
For 100 MHZ, wavelength = 300,000,000 m/s / 100,000,000 Hz = 3 meters.
For 1000 kHz, wavelength = 300,000,000 m/s / 1,000,000 Hz = 300 meters.
For 10 GHz, wavelength = 3×10^8 / 10×10^9 = .03 meters = 3 centimeters.

3. As the energy is directly proportional to the frequency, or inversely to the wavelength, the longest waves carry the least energy. Thus the 100 MHZ wave carries 100x more energy than the 1000 kHz, the 10 GHz carries 10,000x more energy than the 1000 kHz, and 100x more than the 100 MHZ.

4. As kinetic energy is proportional to v^2, the car will have 3^2 = 9x more energy at 30 km/hr than at 10 km/hr., and 6^2 = 36x more energy at 60 km/hr.

5. The hydrogen alpha line is at 6,563 A, so to find its energy, use E = h x f = h x c/wavelength = 6.63×10^{-34} x 3×10^8 / 6.563×10^{-7} m = 3×10^{-19} joules.

6. Picking 5500 A as the middle of the visible spectrum, then each photon of light carries about 3.5×10^{-19} J; a 100 watt light is producing 100 joules per second, or 100 J/s / 3.5×10^{19} J = 2.85×10^{20} photons every second.

CHAPTER 6

Telescopes and Our Insight to into the Cosmos

CHAPTER OUTLINE

Central Question: How do astronomers collect the light from celestial objects?

6.1   Observations and Models

6.2   Visible Astronomy: Optical Telescopes
        A. The basis of optics
        B. Optics and images
        C. Telescopes
        D. Functions of a telescope
        E. Next generation of telescopes

6.3   Invisible Astronomy
        A. Ground-based radio
        B. Resolving power and radio interferometers
        C. Ground-based infrared
        D. Space Astronomy
        E. New views of the sky

6.4   Image Collection and Processing
        A. Understanding intensity maps
        B. Photography
        C. Charge-Coupled Devices (CCDs)

Enrichment Focus 6.1: Properties of Telescopes

CHAPTER OVERVIEW
Teacher's Notes
        The emphasis in this chapter is on the influence of observations on our view of the cosmos. It points out that in many cases dramatic changes in our ideas on the nature of celestial bodies followed observation of these objects, which in turn came about through the development of new instrumentation or technology. Don't get lost in the details of how telescopes work, but concentrate on why they are used.
        If students have been doing unaided-eye observations up to this point (constellations, planet positions, etc.) this would be a good time to begin telescopic observations. Looking at the sky through binoculars or telescopes for the first time will amplify points made by this chapter in a concrete way.

Some astronomers would disagree with the point made that "...astronomy, in comparison to physics or chemistry, is not an experimental science...." It is true that we cannot bring the stars or planets into a laboratory, but scientists (astronomers, astrophysicists--call them what you will) do simulate extraterrestrial conditions in the laboratory to study planetary atmosphere, radiative transfer, gas dynamics, atomic collision processes, and so forth. Additionally, studies of processes within the earth's atmosphere may be considered as astronomy. I remember one professor who always argued that "astronomy begins at the earth's surface"--Reif. I (Michael Zeilik) would be willing to argue that it begins at the earth's core.

Three areas of research and development that are just beginning to show promise of expanding our observing range are spectral interferometry, computer reconstruction of images, and deformable mirrors. Such techniques enable astronomers to reach higher resolution by correcting for atmospheric seeing effects, and have led to the resolution of stellar disks. The use of computers really has reached a crucial stage in observations. Computers control telescopes and their instruments, and so provide intelligent and flexible observing in real time. They can control telescopes remotely, whether in space or on the ground. And computers are the essential toll of image processing.

Be sure your students have a general notion of how to read contour maps after they have studied this chapter. Many subsequent observations in this book are in the form of such maps.

6.1 Observations and Models

In many cases, models predicted phenomena that could not yet be observed. Ptolemy's geometric model with epicycles predicted that Venus must always lie between us and the sun, so telescopically it must be either new or a slender crescent. When Galileo found Venus showed the whole cycle of phases, this proved Venus orbited the sun, not us, a clear break with Ptolemy. The telescope is an experimental tool, for it provides us with observations of new objects, some predicted (neutron stars, black holes), others quite unexpected (pulsars, quasars, x-ray burstars). These observations of course in turn are the basis for new models to be constructed later.

6.2 Visible Astronomy: Optical Telescopes

Zeilik briefly refers to the color distortion of simple refractors (chromatic aberration) on page 110, but it may deserve more attention. Because blue light photons are slowed down more than those of longer waves when they pass through glass, a ray of blue light bends (refracts) more and comes to a shorter focal point; if you viewed a star through the eyepiece of a simple lens refractor, and focused on the red image of a star, the blue light would form a colorful but distracting haze around the red point of light. Focus in a bit, and the blue image sharpens, but now the red gets fuzzy. The shorter the focal ratio of the lens, the worse this problem is, so some early observers, striving for image quality, built exceptionally long refractors (f/100 or greater) to minimize this color distortion. These telescopes were over 30 feet long and had their objectives and eyepieces mounted on the opposite ends of long poles instead of inside tubes. Such clumsy instruments were hard to mount, guide, and look through, for their fields of view were tiny and the image vibrated with every gust of wind. Still, dedicated observers like Huygens used them to find the rings of Saturn; then Cassini used his to note the gap the rings named in his honor.

Today such simple lenses are often used for cheap telescopes, binoculars (field glasses), and small microscopes; most "toys" under $50 (and quite a few deceptive ones more highly priced) do not use more expensive compound (or achromatic) lenses to correct this problem. If you get a small telescope or microscope home, and find even when sharply focused, the images remain colorful and fuzzy, the problem is chromatic aberration; the solution--take the instrument back and look for optics with achromatic objectives and eyepieces. You will pay more, but the viewing will be far sharper.

Newton's solution was to avoid refraction entirely; he used a concave front surface mirror as his objective in his reflecting telescope. As all light waves reflect the same off the mirror, the problem of chromatic aberration is averted entirely. Today reflectors are the major form of astronomical telescopes used in research.

But the spyglass did not die. When English optician John Dolland found how to combine two different glasses into a compound (achromatic) lens about two centuries ago, it was a big hit; small (60-80mm) achromatic refractors are still very popular today, and priced between $100 and $300, are not bad starter telescopes IF provided with good low power eyepieces. The achromatic doublet cures the worst color distortion by using two different types of glass (usually crown and flint) for the objective. The second lens thus "cleans up" the color distortion introduced by the first element fairly well. But in considering larger telescopes, the grinding of such an achromatic doublet requires figuring four optical surfaces, while a primary mirror of a reflector is only figured on the front surface. The glass for such doublets must be optically perfect, hard to achieve as the castings grow in size; a mirror can have imbedded bubbles and striations in the blank what never effect its front surface and image quality. In the range of over about 80mm, the reflectors thus are much easier to build and much cheaper to purchase. The 40" achromat at Yerkes will probably be the largest successful refractor ever built, for even it has the lens sagging under its own great weight. About the turn of the century, the French tried to build a 50" lens, but its images were inferior and it was an utter failure.

As your students first look at the moon through a telescope, they will probably be surprised to note the image in astronomical telescopes in inverted--things look upside down. this is apparent in the diagram on page 109, but not explained in the text. Note the light passing through the bottom of the front (objective) lens reaches the top of the eyepiece, for the eyepiece is located behind the lens' focal point. Technically, this is not the design used by Galileo; his scopes used concave lenses for eyepieces, and he located these in front of the objective's focal point; Galilean telescopes (such as "opera glasses") give erect images. But the field of view for this arrangement is tiny, and Kepler noted that the use of convex lenses for eyepieces, as shown on page 109, gave a better image with a much wider field of view. As astronomers wanted to see as much of the universe as quickly as possible, even if it was upside down, Kepler won.

Yet you are aware that binoculars to give erect images. They do this by passing the light through a set of porro prisms before it reaches the eyepiece, thus inverting the already inverted image (here, at least, two wrongs do make a right); this accounts for the "broken back" design of most binoculars. However, as some light is lost in the prisms, and some color and image distortion is also inevitable, most astronomers still prefer their images sharper and brighter, even if the images are inverted. And of course with video astronomy, the image can be electronically erected: in any event, all lenses used with video cameras give inverted images we never notice.

42

A good pair of binoculars is a fine starting instrument for astronomy, and probably the best value for your money. If your interest in astronomy were to wane, they are always good for birding, football games, or nature study. Just how well would a quality pair of 7x50 binocs for about $40 or so rate, in Zeilik's criteria for telescopes?

First, the 50mm objective diameter is seven times larger than your eye's exit pupil of about 7 mm, so the objective's surface area is 7 squared, or almost 50X larger than your eye. Think of a telescope as a light bucket, whose job it is to gather in as many photons of light as possible; this binocular lets you see stars fifty times fainter than your eye alone could see.

Next, consider the resolution. The 50mm diameter is seven times broader than your exit pupil, so its resolution is seven times sharper. Using the formula on page 109, we find the 50mm (5 cm) objective could theoretically yield resolutions down to about 3" of arc, or show us a crater on the moon only three miles across (not bad from a quarter of a million miles away!). This is much better than your eye alone--most of us have trouble seeing Copernicus (sixty miles wide) on the moon with unaided vision.

Lastly, the 7X means the moon appears magnified seven times, or seven times closer to us, only 35,000 miles distant. Jupiter appear close enough to glimpse its four planet sized Galilean moons, and the Milky Way jumps out at us with star clouds, clusters, and nebulae.

6.3   Invisible Astronomy

As mentioned in Chapter 5, in my 45 year life the new technologies have opened up the entire cosmic symphony to us, far beyond the visual six notes we knew of in 1948. With the Digistar at Pensacola Junior College Science & Space Theatre, we can project images of the entire sky seen not just in visible, but also in gamma ray, X-ray, infrared, and radio telescopes. In our production of Timothy Ferris' "Galaxies", the Digistar starts with the visible Milky Way stretching overhead, then quickly switches to each of the other wavelength ranges. You immediately notice that no single object is visible from one wavelength range to the next! This is evident as well in the text on page 120, but it certainly is more graphic under a forty foot dome.

We love the stars, for they make what we see best--visible light. This is not accident, for our sight is tuned to our star. The sun's yellow color lies in the middle of the visible range of wavelengths. but there is far more to the universe than just stars--if Vera Rubin's "dark matter' model is right, they make up only 10 % of the mass of the cosmos. We must turn to these other energies to really find the nature and fate of the cosmos.

Zeilik does a good job of pointing out just how large radio telescopes must be to get good resolution. If your students have seen the movie, "2010", remind them of the plans made for the Leonov mission between the Russian and American scientists on the stairs of just one of the 85 foot wide radio dishes in the VLA. Also encourage your students to read more about the plans for new telescopes, some already under construction, described in the articles from Astronomy.

6.4   Image Collection and Processing

In considering why CCD's are replacing photography, note how easy it is to manipulate the video image off of a VHS tape (changing the brightness, contrast, or tint on the TV, for instance). Now compare this with the tedious, expensive, and time consuming job to retaking a print from an original negative in the darkroom.

Video astronomy has indeed come of age. With the video image, you see results instantly, as the telescope is focused on the object, and CCDs can detect far fainter objects in less time. Also, you can relay your electronic data to your colleagues--who does research alone these days, anyway!

Practically, as CCDs gather data much faster than film can, the same expensive telescope can be used to gather far more information in a single night. Hence the 200" at Palomar can not be used to take photos any more, and the 400" Keck will never take a photo. Photo emulsions can capture only about 10 % of the photons striking them, but the .2 lux CCD in your new videocam may be about 90 % efficient in turning photons into an electronic image; these systems are available for about $300 now, and your telescope should be set up to give your students video images on TV monitors.

ADDITIONAL RESOURCES

Transparencies 6.4, 6.6, 6.7, 6.9ab, 6.14, and 6.25abc as well as slides 6.12b and 6.13d will be of value in class presentations.

If you can, find Philip Morrison's first episode of "The Ring of Truth". It did a great job of showing just how large a change in human perception the telescope brought about. The term "remote sensing" will have a whole new meaning to your class. Check your local PBS station for its availability.

If your planetarium has a Digistar projector, try to get Timothy Ferris' "Galaxies" from the Hansen Planetarium in Salt Lake City. Its a fine program for Unit IV, and the different views of the sky in various wavelengths certainly fit in well here.

While it is dated, the old Nova episode, "Beyond the Milky Way" is a neat historical document on how the new technologies of a decade ago were opening up the universe even then.

SUPPLEMENTARY ARTICLES

1. "Balloon Borne Astronomy," Astronomy, August 1984, page 6.
2. "Photoelectric Photometry--to better measure starlight," Astronomy, June 1986, page 74.
3. "Building your own simple telescope," Astronomy, January 1987, page 63.
4. "Shielding the Night Sky," Astronomy, September 1988, page 47.
5. "Easy Astrophotos," Astronomy, September 1989, page 70.
6. "Seeing Sharp," Astronomy, July 1990, page 38.
7. "Video Astronomy comes of age," Astronomy, August 1990, page 60.
8. "New Telescopes for the Future," Astronomy, November 1990, page 34.
9. "First Light on the Keck 400" Telescope," Astronomy, April 1991, page 42.
10. "ROSAT's Penetrating X-Ray Vision," Astronomy, June 1991, page 42.
11. "Seeing the universe in the infrared," Astronomy, June 1991, page 50.
12. "GRO: The Violent Universe in Gamma Rays," Astronomy, July 1991, page 44.
13. "Using the Hubble Space Telescope," Astronomy, December 1991, page 43.
14. "Three Nights on Kitt Peak," Astronomy, April 1992, page 38.
15. "The Birth of Radio Astronomy," Astronomy, June 1992, page 46.
16. "First Light for the Keck 400" Telescope," Astronomy, August 1992, page 22.

17. "Fixing Hubble's Vision," <u>Astronomy</u>, February 1993, page 22.
18. "Big Scopes for the 21st Century," <u>Astronomy</u>, August 1993, page 48.
19. "The Best Cameras for Astrophotography," <u>Astronomy</u>, September 1993, page 74.
20. "Putting Hubble Right," <u>Astronomy</u>, March 1994, page 24.
21. "Virtual Sky (CCD imaging)," <u>Astronomy</u>, March 1994, page 70.
22. "Hubble Better than New," <u>Astronomy</u>, April 1994, page 44.
23. "100 Years on Mars Hill (Lowell Observatory)," <u>Astronomy</u>, June 1994, page 28.
24. "Telescopes that Fly," <u>Astronomy</u>, November 1994, page 46.
25. "T is for Telescope," <u>Astronomy</u>, November 1994, page 70.
26. "Astronomy's Future in an Age of Budget Cutting," <u>Astronomy</u>, July 1995, page 40.
27. "Catch a Comet on Film," <u>Astronomy</u>, January 1996, page 86.
28. "Easy Astrophotography," <u>Astronomy</u>, February 1996, page 74.
29. "Choosing a Camera for Astrophotography," <u>Astronomy</u>, March 1996, page 76.
30. "Journey to the Outer Limits (the TAU mission)," <u>Astronomy</u>, August 1996, page 36.
31. "Lunar Windows to the Heavens," <u>Astronomy</u>, September 1996, page 50.

ADDITIONAL DISCUSSION TOPICS

Building Cheap Refractor Telescopes

Back in March 1983, Merry and I built the first copier lens telescopes and publicized them in the journal of the Escambia Amateur Astronomers, the <u>Meteor</u>. They caught on, and soon many variations on the theme appeared; we tried to show the diversity of such designs in our Winter 1986 issue of <u>New Horizons</u>, the quarterly newsletter of the Southeastern Region of the Astronomical League; we still get requests for copies of this article. Since then many popular surplus eyepieces have sold out, and some copier lenses are also no longer available. What is still available now?

Much work in promoting the construction of cheap, creative lens telescopes has been done during the past several years by Rico Tyler of Franklin Simpson High School in Franklin, Kentucky. His Project Spica has built dozens of simple refractors for student use; one was even presented to the Governor of Kentucky and is on display in his office. You can contact Rico at (502) 586-3273, or write him c/o Project Spica, Franklin Simpson H. S., Franklin, KY 42134.

To build any telescope, you must have: 1) a large lens or mirror designed to collect and correctly focus the light, 2) a second set of lenses, or eyepiece, to magnify and focus the light for your eye, and 3) a tube to hold both lens elements properly aligned, with provision for focusing the eyepiece for different objects and individual observers. A helpful accessory is to add a star diagonal, allowing the eyepiece to be tilted 90 degrees for more convenient viewing when objects are near the zenith, or looking at objects on Earth instead of in the sky. You must also have a mount capable to holding the telescope stably at high magnification, yet also capable to being pointed smoothly to different objects and allowing you to track the objects across the sky easily as the Earth rotates.

You can of course buy such telescopes in department stores for about $70-$130, but for a fraction of that price you can assemble the parts and build your own simple refractor (lens type) telescope that will preform better and teach you something about optics in the process.

First, where can you get cheap, high quality optics? Three good sources of supply are:
* C&H Sales, P.O. Box 5356, Pasadena, CA 91107 phone 1-(800)-325-9465
* American Science and Surplus, POB48838, Niles, IL 60714-0838 phone (708) 475-8440
* Sky Instruments, MPO Box 3164, Vancouver, BC Canada V6B 3Y6 (604) 270-2813

Write or call each for their current catalog. Look through the listings and find what is currently available at a good price. As of September 1996, I can suggest the following as good buys in objectives and eyepieces:

* C&H has a Bausch and Lomb copy lens with an 8.25" (210 mm) focal length, and a focal ratio of f/4.5, with click stops at f/5.6, f/6.3 and f/8. Normally you would want to use the full aperture of the lens, and leave it at the most open setting (f/4.5) but with some bright objects like the Moon, Venus, and Jupiter, stopping the lens down might help the image quality. This is item OL3301, and costs $12.50; slightly more expensive versions of the same basic lens with more click stops are not worth the extra cost.

* C&H has some high quality microfilm reader lenses from Olympus for eyepieces for only $6 that work quite nicely; order both the 14.2 mm (OL8958) and 31.9 mm (OL8952) lenses at this great price while they are still available.

To determine how much magnification an objective and eyepiece give you, divide the objective focal length by the eyepieces; for the scope here, you would get 15x from the 14.2 mm eyepiece with the 210mm objective, and about 7x from the longer f.l. eyepiece.

* American Science lists a good copier lenses, #20650 for $12.50. This giant from Tominon was used in IBM copiers, and has a 230mm focal length at f/4.5. The lens barrel is 3.7" in diameter, so you will probably use 4" PVC to hold it.

* American Science Center has a good low power eyepiece, #22117, for only $2.50. This Tominon lens has a focal length of 26.9 mm, and has a barrel of 23mm diameter, just fitting into Japanese size eyepiece holders; a few wrappings of masking tape and a Kodak film canister will get it up to 1.25" standard. You might also have fun building your own eyepieces; the small lenses kit, #3707, is only $2.00, and combined with the achromat kit, #3715, for $3.00, you may be able to build some surprisingly good eyepieces for a song; again, use the Kodak canisters for eyepiece tubes, and odd pieces of plastic for bushings inside.

* Sky Instruments stocks achromatic (color corrected) objective lenses of 60mm aperture in both 700 mm (6F12) and 900 mm (6F15) focal lengths; these are $29.95, and probably worth the extra cost in terms of improved performance, even though the longer tubes required for the longer focal lengths may impair portability; if this is important, the 700 mm version is best.

For the tube, PVC plastic works well; it comes in a variety of diameters, can be easily sawed, drilled, and machined, and is frequently free if you visit a construction site where odd lengths are being discarded by the plumbers. If not, most hardware stores keep a good selection in stock at low prices. To mount it on a photo tripod, you need only find the center of balance of the finished tube assembly, and drill and tap a 1/4"x20 hole into that point on the tube, then screw it onto your tripod; be sure not to strip the threads in the PVC, and check to insure the assembly is secured in place well. You could of course make a wooded saddle to hold your tube, with a 1/4x20 tee nut (countersunk from the top of the center board of the cradle) and a hose clamp to secure the tube to the saddle assembly.

Try to find a tube just large enough to allow the objective to slip inside; use several wrappings of masking tape for shim if needed. For the 60mm objective, 2.5" pvc should work well. To insure the objective does not slip in the tube, you could drill in little set screws just fore and aft of the intended objective position (removing the lens first, of course!).

Be sure it is exactly perpendicular to the axis of the tube by pointing the tube assembly (no eyepieces) at the Sun and noticing the projected image and shadow of the tube on a sheet of white cardboard; if properly aligned, the bright point will be centered in the round shadow of the pvc tube.

How long does the tube need to be? It will be a little less than the focal length of the objective; for the B&L lens, 7" long will be about right, while the 700 mm will need a much longer one of about 25". Remember, you will need a focusing drawtube to hold the eyepiece, and it will slide out far enough to get the eyepiece to focus. If you plan to construct a star diagonal, cut at least another inch off the length of the tube. If needed, you can easily cut more off the eyepiece end of the tube, or add a smaller diameter drawtube to the eyepiece end to adjust for the correct focus. To fit these drawtubes into the outer tube, I suggest several wrappings of wide (1.5-2") masking tape to build bushings for a snug fit, yet loose enough for slide focusing where needed. Cardboard tubing also works here, provided you keep it dry (contact paper is an ideal covering for both PVC (which does not paint well) and cardboard (ditto).

As far as mounting the eyepiece, you may be able to find pvc bushing (2" to 1") that will fit the inside of the outer tube and size down to a 1" (really closer to 1.25" aperture) to hold the eyepiece and star diagonal, if desired. If not, use two smaller tubes for drawtubes to get down to the desired 1.25" size to hold the eyepiece.

You can buy commercial eyepieces with 1.25" metal tubes from many telescope suppliers for $30 and up--check out listings in Sky and Telescope and Astronomy. A 25mm kellner type will work well as an all-purpose starter ocular. If you decide to go with cheaper microfilm lenses, realize they have narrower fields of view, and you must mount them in a 35mm film canister (cutting off the bottom end, of course) with a bushing of masking tape (3/4" or 1" usually works best).

If you still need a bushing to hold the canister in the focuser, start by wrapping the 1" tape, sticky side out, around the outside of the canister at least two full rounds. Then tear it, and finish your bushing with the sticky side in on top of the first two rounds; this new tape tube will be your fine focusing drawtube for the eyepiece. You can use the same tactic for adapting Japanese standard (.965") eyepieces to 1.25" format.

These little scopes are a blast to build, even for kids. My son Michael built a very nice one in about 30 minutes when he was still eight years old, and he takes pride in showing it to other kids at out club stargazes. The view of the Moon is fantastic for a $30 instrument, and it will show the phases of Venus, four large moons of Jupiter, rings of Saturn and its large moon titan, pick up both Uranus and Neptune, all Messier objects, etc. Its projected image of the Sun reveals granulation as well as tiny sunspots. As a 400mm telephoto lens, it give great slides of the prominences in the 1991 total solar eclipse. It is small enough to keep with us in the car at all times, along with a photo tripod. For Astronomy Day 1993, my six year old son Trevor put together the B&L scope below in about 10 minutes. We helped build twelve of these in April 1996 with a class of fifth graders at Brownsville Science Magnet School in Pensacola. If you haven't built a small refractor like this, it is a great class project and the result will be a scope you will all enjoy and use for many years to come. They can be built in about one fifty minute class period, and are also very useful as "macroscopes", for the sliding tube can be pulled out to focus on objects just a few feet away--an ant hill looks like a Japanese SF movie! Remind your students that the farther away an object they are focusing on is, the more they need to push the eyepiece in toward it; as you focus on closer objects, pull the eyepiece back out toward you.

The Telescope Mounting Revolution

While a camera tripod works well for a low-power telescope like the copyscopes described above, at high magnification you need a more stable mount that does not shake in the breeze and move smoothly. For astronomical observing, the situation is even more complicated. To compensate for the earth's rotation, a telescope needs to be capable of tracking diurnal motion across the sky in circles centered on the pole, and also be driven by electric motor at the rate of one earth rotation per day; the sidereal rate is 23 hours, 56 minutes. This means either the telescope mount must be inclined at an angle to match your latitude (the "equatorial" mount) or that both axis of the traditional alt-azimuth mount (such as a camera tripod) must be driven horizontally and vertically simultaneously and at varying speeds as the sky position changes.

During most of this century, a variety of equatorial designs have been tried. The fork and wedge is popular with Schmidt-Cassegrains like those from Meade and Celestron, while the German design remains the standard for commercial newtonians. But these are heavy, and expensive to build and technically difficult to fabricate in large sizes for professional telescopes. The huge horseshoe for the Palomar 200" was a technological wonder, but its great cost and weight pushed technology of materials and hydraulics to the limit, making people doubt if any larger instrument could ever be successfully built.

Then in the 1960's the Russians built a new 236" reflector, mounting it on the much simpler and lighter alt-azimuth mount. Using gravity as a friend instead an enemy, the 236" was much less expensive and more compact than Palomar. While the thick main mirror of the 236" has had thermal problems and not lived up to expectations, the mounting was a great success. The Russians secret was using computers (no more sophisticated than Commodore 64's) to control servomotors driving both axes simultaneously. The corrections for the object's changing positions in the sky, even the changing refraction due to the atmosphere itself, were all computed hundreds of times a second. In effect, equatorial tracking was recreated at a fraction of the cost.

Soon all new telescopes were being built this way, including the 400" Keck in Hawaii. Even amateur instruments are coming around. In the 1960's John Dobson of the San Francisco Sidewalk Astronomers introduced "dobsonian reflectors," using thin, home-ground mirrors mounted in light, portable wooded alt-azimuth cradles, to make large, cheap, easily transportable light buckets; I personally have helped build several, including a 12" for our local club. With great light grasp yet smooth, easy motions atop teflon bearings, they were immediate hits at star parties as people marveled at spiral galaxies and globular clusters as never seen visually before. But they could not be used for astrophotography, and required constant hand correction.

Now the cost of servomotors and PCS has dropped until these dobsonians can be made to track equatorially as well. A new commercial model from Meade doesn't even come with a wedge as standard equipment; its alt-azimuth mount works fine with a hand-held control box and a 9 volt battery. All you need to do in setting up the scope is to point it to two stars and identify each to the computer control. The computer then will calculate your exact latitude and the sidereal time; next it takes over controls of the scope, driving the scope to match the sidereal rotation rate and keep the object centered in your eyepiece. If you want a new object, just tell the computer what is wanted, such as Saturn, or M-51, etc, and it will (if the object is above your horizon, at least) slew the scope over to the object's exact position. Certainly takes all the drudgery out of star hopping through your finder scope! See the ads in the latest issues of <u>Astronomy</u> and <u>Sky and Telescope</u> for more info.

The Power of Interferometry

The Very Large Array near Socorro, New Mexico is a striking example of a radio interferometer, where several radio dishes are arrayed to combine their weak signals via computer into a stronger signal with far better resolution than any single dish could give. The increase in detail in the radio images is incredible. This is easiest to do with long waves, were the dishes do not have to be so accurately figured to give signals easier to enhance via computers. But now our optics and computer programs are becoming so advanced that optical interferometers are being assembled.

Early in 1996, twin 16" scopes were used to first optically resolve the close binary Capella, with resolution exceeding even the HST. As with radio dishes, the separation between the twin scopes determined the incredible resolution. Of course, the farther apart the scopes are, the harder it is to process the twin images via computer, but advances are being made. For instance, the Keck 400" in Hawaii now has an identical twin, several hundred feet away from it. While the technology to combined the twin images from the Keck "binoculars" does not exist at this writing, by the time we prepare a ninth edition of this text, I bet it will be in operation. This will be another revolution in our study of stars and even planets orbiting them!

DEMONSTRATIONS AND OBSERVATIONS

The Eye as an Astronomical Sensor

It is true the telescope greatly amplifies the eye's abilities, but the eye is still a superb astronomical instrument. For certain types of observing, it is still unmatched. When dark adapted, it can view over 140 degrees of Milky Way, almost horizon to horizon. From a dark, clear site, you can see the dark nebulae standing out in the foreground, in front of the spiral arms of the Galaxy. One of my favorite mind exercises is to lie back on a blanket with a clear horizon, with noting on planet earth visible at all. Concentrate on the galaxy, and the stars go out of focus; you are star trekking into the heart of the galaxy, like an astronaut on a journey of thousands of years and trillions of miles. On such a summer mind trip, you are soon brought back to reality by the flash of a meteor overhead, another item the eye detects superbly. Our eyes can sweep over a vast portion of the visible sky, and they are very sensitive to sudden changes in brightness, such as the burning of that bit of cosmic debris. The American Meteor Society still relies chiefly on visual reports and plots in determining the radiant point and strength of meteor showers. As these can change greatly from year to year, the need for these observations continues.

A key to successful visual observing is dark adaptation. Your retinas have two types of sensors, cones and rods. The cones lie near the center and optic nerve (blind spot), and a used for normal, full-color vision. They require fairly high light levels to stimulate their color sensors, so on a dark night only a few of the brightest stars show notable color, such as bluish Vega, yellow Capella, orange Arcturus, and reddish Antares and Betelguese. When a telescope objective gathers more light, the view at the eyepiece is more colorful. Particularly double stars of different temperatures may show color, such as the fine orange and blue pair of beta Cygni, easily split at 20X with a copyscope.

The rods do not require nearly so much light, and are relied upon for most observing on dark nights. They are sensitive to only black and white, however, and we need time to shift our eye's chemistry from bright colors to far paler shades of gray. When we first enter a theater, we need to adjust to the darkness, so as not to spill our popcorn and soda on the folks we tripped on in our row.   If you are bringing your students into the planetarium to observe constellations, be sure to schedule time for them to get adjusted to looking for stars, particularly if they are just coming inside from a brightly sunlit day.

Also, the rods are found chiefly around the sides of the retina. Often if you are looking for a faint object, you try starring at it without success. In frustration, you start to glance away-- wait, there it is! You have found averted vision; by looking off the side of the eyepiece field of view, you employ more of the sensitive rods, and can spot fainter objects.

A key difference between the eye and the photographic plate and CCDs is that the eye can not store up photons. The retina erases itself about twenty times a second, so if you do not see something in that span, you will not see it at all. This is necessary, for we must note moving objects that could threaten us, but it also means a 30 second time exposure with fast film will capture far more stars and star colors than your eye can ever see. On the positive side, when observing planets at the eyepiece, a visual observer can spot elusive planetary detail when the atmospheric seeing is stable for a fraction of a second; the time exposure is smeared by the shifting currents and the detail is blurred on the photograph.

Light Pollution and Fighting It

Night is a right! In a world of city dwellers, it is hard to see the stars anymore. Urban air and light pollution makes most constellations unrecognizable to anyone near a major urban area, including the vast majority of our population. If you can take your students to a dark beach or rural area, many will be amazed at the difference in what they can see under truly dark adapted conditions. But progress marches on, crime threatens the streets, and seemingly our planet must grow more brightly lit night after night.

Astronomers should not give up, however. There are BAD lights, and not so bad lights. The worst in every regard are incandescent (screw in tungsten) bulbs. They create a broad continuum of light that can not be filtered out, and most of the electricity goes into heat, not light. Fluorescent lights are far more energy efficient, but the phosphor coating still give a continuum that is hard to filter out. Mercury vapor lights create just a few emission lines, and the dark sky filters you will see advertised from Lumicon and others block most of these emission lines, making the sky through the eyepiece appear much darker without loosing much of the continuum from the stars or nebula emission lines. The best of all is LOW pressure sodium, presently the most efficient way to turn electricity into visible light. As almost all the energy lies in the sodium doublet lines in the yellow, some people do not like the monochromatic color, but it is easy to filter out, and several communities near major observatories have made this type of lighting required for outdoor security purposes, darkening the night sky without loosing any security and also saving thousands of dollars in electric bills. Get your students active in urging your city and power companies to consider this "green and black" solution to lighting.

The organization that is leading the fight against wasted light is the IDA (International Deep Sky Association). Check out the organizational addresses available on <u>Astronomy</u> and <u>Sky and Telescope's</u> home pages and send off for more information on how to help preserve the beauty of the night sky.

ANSWERS TO STUDY EXERCISES

1. Galileo's observations of Jupiter's moons showed the earth was not the center of all celestial motion; they obviously orbited Jupiter, and in accord with Kepler's laws as well. His observations of sunspots showed the sun rotated and changed day by day; the sun was not the perfect, unchangeable body the Greeks imagined. The moon too was imperfect, as shown by its many craters and dark, flat mare. The complete cycle of Venusian phases proved Venus orbited all the way around the sun, not on the epicycle between us and the sun which Ptolemy proposed. The resolution of the Milky Way into stars proved the celestial sphere to be too large to be real.

2. While all of the observations mentioned in #1 above helped support Copernicus, the critical observation of stellar parallax needed a far better scope; it would be the middle of the nineteenth century before telescopes and measuring devices proved the earth was revolving from one side of the sun to the other over a six month interval.

3. For astronomers, the most important telescopic function is to gather light, to reveal a universe far more vast than the eye alone can glimpse.

4. As radio waves carry less energy per photon than do visible light waves, radio dishes must be built far larger to gather enough radio waves for a clear signal and good resolution. The methods of detection are different--radio telescopes do not use photographs or visual detectors, but are really just sensitive radio receivers, mush as photometers just monitor the light intensity. Both radio dishes and reflecting telescopes use a curved reflector to focus the waves they collect, and there are even cassegrain design radio reflectors. As the sun is a weak radio source, radio telescopes are often operated in broad daylight or under cloudy skies (but not during a thunderstorm!). Also, dust does not block radio waves, so we can map our galaxy much better with radio waves than light, where dark nebula hide 90 % of our home galaxy.

5. As the resolution increases as energy increases and wavelength decreases, then the same dish reflector would give best resolution in optical wavelengths, then less in infrared, and the poorest image as a radio dish. Working at 21 cm wavelength, for instance, a radio dish would need to be about 3.5 kilometers (or over two miles) across to match the one arc minute resolution of the human eye in visible wavelengths.

6. Space telescopes like Hubble are not limited by the turbulence of the earth's atmosphere, so the images in space are much clearer and sharper. Without the airglow of the ionosphere and light pollution, the space sky is far darker, so longer exposures will reveal far fainter objects as viewed from space. Without the ozone layer or greenhouse effect, the space telescope does much work in both the infrared and ultraviolet wavelengths we miss under our atmospheric blanket. Also, there are no clouds, so the telescope can be used full time.

7. Infrared radiation is hindered less by interstellar dust; cool objects are brighter in the infrared than in visible light; infrared telescopes can be used in the daytime; and infrared resolution is less affected by atmospheric seeing as visual observations are.

8. X-rays are blocked by our ionosphere (that's why it is ionized), so ground-based scopes will not detect them. Balloon borne scopes did some pioneering work, but most is now done from space.

9. As a truly large reflector comparable to Keck is hard to place in orbit, the best use of Hubble is to take high resolution shots not obtainable by any scopes working through our turbulent blanket of atmosphere. By contrast, long exposures with Keck under dark skies can reveal very faint objects.

10. It light gathering power, so buy the largest one that fits in your budget and observing routine.

ANSWERS TO PROBLEMS & ACTIVITIES

1. The diameter of the telescope is the focal length/f-ratio, or $900/15 = 60$ cm. If the 10 cm telescope had a visual resolving power of 1.4 arc seconds, then the 60 cm scope will give six times better THEORETICAL resolution, or .23 arc seconds (not attainable on earth's surface. As the telescope is a "light bucket", the scope is $60/.5 = 120$x the diameter of your 5mm exit pupil, or has a surface area of $120^2 = 14,400$x larger than your eye, revealing stars 14,400x fainter than your eye.

2. Again, compare their surface areas; $(5m/2.4m)^2 = 2.1^2 = 4.3$x

3. The focal length = f/ratio x diameter = 8" x 10 = 80"

4. Resolution depends upon objective diameter, so as the HST has 24x the diameter of the 10 cm objective, it will resolve 1.4"/24 = .058", achieved and bested by the last repair mission.

5. As resolution depends inversely upon wavelength, the 2 cm radio waves carry much less energy than visible waves (about 5500 A or 5.5×10^{-5} cm); the actual ratio is $2/5.5 \times 10^{-5}$ or about 3.6 million times less energy. To equal the resolution of a 10 cm scope, the radio dish would need to be 36 million cm, or 363 km across. The 40 km VLA thus gets only about 1/9th resolution.

6. Magnification = objective focal length / eyepiece focal length, so the scope must have an objective focal length of 150 x 20 = 3,000 mm. To get 75x, your need a 3,000 / 75x = 40 mm eyepiece; for 250x, you need 3,000 / 250x = 12mm eyepiece.

7. As the naked eye can resolve about an arc minute, and the moon is about 400,000 km distant at apogee, then the smallest features are sin .0167 deg = x / 400,000 km , so x = 120 km or about 70 miles across; near perigee, features like Copernicus can be noted by sharp eye.

8. Again, we can resolve about one arc minute, so sin .0167 deg = 1 km / x; thus x = 60 kilometers away.

9. In optical wavelengths, the Raleigh criterion is 14" per cm of aperture. As the interferometer had a diameter of 10 km = 1 million cm, its theoretical resolution is .000014 " of arc. As our planet has a disk of 12,000 km diameter, then sin .000014" = 12,000 km / x; x = 50 billion km or 3,300 AU.

CHAPTER 7

Einstein and the Evolving Universe

CHAPTER OUTLINE

Central Question: How did Einstein's ideas about gravitation lead to an evolving model of the cosmos?

7.1   Natural Motion Reexamined
        A. Newton's Assumptions
        B. Motion and Geometry

7.2   The Rise of Relativity
        A. Mass and Energy, Space and Time
        B. The General Theory of Relativity
        C. The Principle of Equivalence
        D. Weightlessness and Natural Motion

7.3   The Geometry of Spacetime
        A. Euclidean Geometry
        B. Non-Euclidean Geometry
        C. Local Geometry and Gravity
        D. The Curvature of Spacetime
        E. Spacetime Curvature in the Solar System
        F. Experimental Tests of General Relativity

7.4   Geometry and the Universe
        A. Cosmic Geometry
        B. A Quick Tour of the Universe
        C. The Expanding Universe
        D. Hubble's Law
        E. The Meaning of Hubble's Law

7.5   Relativity and the Cosmos
        A. Escape Speed and the Critical Density
        B. The Future of the Universe

Enrichment Focus 7.1: Hubble's Constant and the Age of the Universe

CHAPTER OVERVIEW
Teacher's Notes

Since the main theme of this chapter is Einstein's geometrical concept of gravitation, it is perhaps best to stick close to the geometry and observations that support the general view of an expanding, evolving universe. Details of specific cosmological models are dealt with in chapter 21. Note this chapter deals with dynamical cosmology: the expansion of the universe. It does not cope with physical cosmology, which focuses on the evolution of the physical conditions of the universe and its contents. This topic will be dealt with in chapter 21. Also not that this chapter continues the central theme of chapters 2, 3, and 4; the nature of natural motion and its influence on cosmological models.

Many students will experience some difficulty in comprehending the geometrical concepts and the significance of the observations, since we line in a small, Euclidean, nonrelativistic region of spacetime. Again, physical models which can be demonstrated or manipulated will ease comprehension. If possible, use a spherical surface, for instance a slated globe, to illustrate difference between Euclidean and non-Euclidean geometries.

One demonstration that may be useful is to use a rubber balloon (approximately spherical) as a two dimensional model for the expanding universe. In this model, spatial dimensions are confined to the surface of the balloon. Dots may be drawn on the balloon, and the balloon then inflated. Measuring distances between dots at successive stages of inflation (analogous to different times) yields a "Hubble relation," and also demonstrates that the same result is obtained regardless of the place chosen as the reference point for measurement. Warning: Students may be mislead by this analogy to infer that the universe has an edge. ("The Edge of the Universe..." that famous, frequency, and false phrase from science fiction.) Point out a rubber rope with "galaxies" attached that can simulate a one-dimensional "line-land" universe. Again, uniform expansion results in a Hubble law.

7.1 Natural Motion Reexamined

A point worth noting with this section is that to date we have not need to use Einstein's work in flying around the solar system. Newton's laws have proven sufficient to guide the Voyagers on their Grand Tour, sweep Ulysses over the sun's poles, and send Galileo to Jupiter orbit. Even when Voyager II was accelerated past Neptune in 1989, it final speed of about 120,000 mph was still only .02% of the speed of light, far to slow to require any relativistic correction. New propulsion technologies (ion drive, mass drivers, nuclear propulsion?) will be needed before we can consider reaching the stars in reasonable lengths of time, and thus need to apply Einstein's work in making our flight plans.

7.2 The Rise of Relativity

One of my favorite examples of $E = mc^2$ is the piece of antichalk. Most students know that the power supply for the Enterprise on Star Trek is matter-antimatter annihilation; the dilithium crystals are the catalysts which control the rate of energy production. As they would react with any material object, the antimatter must be confined via a strong magnetic "bottle" to avoid blowing up the starship.

In class discussion, imagine a small piece of chalk (about 5 grams, the mass of two pennies) meeting an equally massive piece of antichalk. The resulting explosion would convert ten grams of matter-antimatter into sheer energy, a total of 9×10^{14} Joules. If we had a means of hooking it into our power grid, this is equal to about 250 million kilowatt hours!

Call the students' attention to the free fall time through the earth's center on p. 131. Note how closely it matches the 90 minute orbital period of satellites in LEO (low earth orbit), like the Space Shuttle.

7.3 *The Geometry of Spacetime*

Episode 10 of "Cosmos" reinforces well the complex geometrical considerations discussed here, particularly the problems of dealing with multi-dimensional space and time.

In considering geometries, it might be good to use the analogy of comet orbits around the sun. Short period comets like Halley's are operating in a closed (spherical Oort Cloud) system. Comets with parabolic orbits are in a flat geometry, while those that can escape the solar system entirely are in hyperbolic trajectories. Note that it is the balance between the gravity of the entire solar system and the kinetic energy of the comet that determines its fate. This concept is enhanced by the discussion of the curvature of spacetime by the sun on page 134.

In considering the gravitational lensing of the stars' images on page 135, be sure the students note the star's images are displaced outward, away from the massive body. Later in our discussion of quasars in Unit Four, we will find many cases where the massive galaxy seemingly is sitting directly in front of a far smaller and more distant quasar. Normal geometry would have us expect the galaxy would completely hide quasar, yet the gravitational lens bends the bright quasar's light into multiple images, and may even focus and brighten the quasar's images for us.

7.4 *Geometry and the Universe*

If students wonder at this point why the value of H is so ill-defined, note that it must be experimentally found by plotting the recession velocities versus the estimated distances to many galaxies. While the recession velocities are easily found via doppler shifts of spectral lines, the distances to galaxies are much harder to measure precisely. To find these distances of millions or even billions of light years, we need indicators that are both predictable and very luminous. A sad rule of the cosmos is that the more energetic a body is, the LESS predictable its behavior will be! We try using a whole sequence of luminous yardsticks, with mixed results. Indicators include cepheid variables, supergiants, sizes of planetary nebulae, H II regions, and globular clusters, supernovae, and even the sizes of the largest spiral galaxies. Yet these indicators often give conflicting results on the distance to the same galaxy. The greater accuracy of the improved Hubble Space Telescope has given more precise and reliable data, for it can spot the less luminous and hopefully more reliable indicators at far larger distances. But also consider that the universe may not be expanding uniformly; it is possible that denser regions, such as our own Virgo Supercluster of galaxies, are slowed down by greater local gravity, giving a lower value of H for objects within a hundred million light years.

In considering the Hubble constant and the universe's age, my favorite analogy is the movie projector running backwards. The faster the projector runs backwards, the less time it takes to get back to the beginning of the movie, the "Big Bang" in this case.

7.5 *Relativity and the Cosmos*

One of my favorite examples of Einstein's spacetime comes from the closed model of the cosmos. It is easy to visualize reversibility in three dimensions; I can easily go left or right, forward or back, up or down. But in Einstein's fourth dimension, such reversibility is not evident. Tomorrow always follows today; you never (unless you are Bill Murray in "Groundhog Day") wake up and find it is yesterday again. Time may flow in only one direction because the universe is larger today than it was yesterday; as long as this is true, time passes normally.

But the concept of a closed cosmos brings up the possibility that radiation's triumphant expansion of the Big Bang is a temporary victory. If billions of years from now, the greater than critical density of the cosmos slows down, then stops the expansion, the following "Big Crunch" will cause distant galaxies to blue shift in toward us and the cosmos to collapse in upon itself. Perhaps as Sagan hints in Cosmos 110, "The Edge of Forever", even causality will reverse. In our expanding cosmos, effect follows cause. In a contracting cosmos, your students might start with the final exam, and end up on the first day of class, perhaps an appealing concept to some.

ADDITIONAL RESOURCES

Transparencies 7.4, 7.5ab, 7.6abc, 7.8ab, 7.10ab, 7.11ab, and 7.15abc, as well as slide 7.13 will help your class presentations, as will Episode Ten of <u>Cosmos</u>, "The Edge of Forever."

SUPPLEMENTARY ARTICLES

1. "Matter and Evolution in the Universe," <u>Astronomy</u>, September 1984, page 67.
2. "Discovery of the Expanding Universe," <u>Astronomy</u>, February 1985, page 18.
3. "The Fate of the Universe," <u>Astronomy</u>, January 1986, page 6.
4. "To the Big Bang and Beyond," <u>Astronomy</u>, May 1987, page 87.
5. "The Cradle of Creation," <u>Astronomy</u>, February 1988, page 40.
6. "The Search for Dark Matter," <u>Astronomy</u>, March 1988, page 18.
7. "The Structure of the Visible Universe," <u>Astronomy</u>, April 1988, page 42.
8. "Recreating the Universe," <u>Astronomy</u>, May 1988, page 42.
9. "How far to the Galaxies?," <u>Astronomy</u>, June 1989, page 48.
10. "The Legacy of Edwin Hubble," <u>Astronomy</u>, December 1989, page 38.
11. "Supercomputing the Universe," <u>Astronomy</u>, December 1989, page 20.
12. "Is Cosmology a Sometime Thing?," <u>Astronomy</u>, July 1991, page 38.
13. "Shedding Light on Dark Matter," <u>Astronomy</u>, February 1992, page 44.
14. "Beyond the Big Bang," <u>Astronomy</u>, April 1992, page 30.
15. "COBE's Big Bang," <u>Astronomy</u>, August 1992, page 42.
16. "Counting to the Edge of the Universe," <u>Astronomy</u>, April 1993, page 38.
17. "A New Map of the Universe," <u>Astronomy</u>, April 1993, page 44.
18. "The Great Attractor," <u>Astronomy</u>, July 1993, page 40.
19. "In the Beginning," <u>Astronomy</u>, October 1993, page 40.
20. "Everything You Wanted to Know About the Big Bang," <u>Astronomy</u>, January 1994, page 28.
21. "How Old is the Universe?," <u>Astronomy</u>, October 1995, page 42.
22. "Wormholes and the Spacetime Paradoxes," <u>Astronomy</u>, February 1996, page 52.
23. "What Happened Before the Big Bang?," <u>Astronomy</u>, May 1996, page 34.

ADDITIONAL DISCUSSION TOPICS

The Strong Nuclear Force

The strong nuclear force is compared by Sagan in Cosmos 109 to a pair of hooked fingers. As long as the temperature is too low, the protons can not get close enough for the strong force to work, and the electromagnetic repulsion keeps the protons apart. But at about 10 million K, they collide violently enough to get them within the strong force radius, and the fingers hook, uniting the protons in nuclear fusion. Without this binding force, the ONLY element in creation would be hydrogen, making the universe a rather dull place!

To comprehend the scale of the strong force, ask your students what units the yield of nuclear weapons is measured in. Most will respond megatons. Millions of tons of what, you ask? Some will know TNT is the standard explosive used. It is no great military secret that a megaton yield thermonuclear device weighs about a ton; based on this, you ask, how much more efficient in turning matter into energy is the strong force than the normal electromagnetic forces in everyday chemical reactions. They can see that from the same weight of warhead, the hydrogen bomb releases a million times more energy (megaton) than does a ton of TNT.

The Search for Vulcan

Zeilik briefly notes the search for the hypothetical planet Vulcan in his discussion of Mercury's orbit and Einstein's relativity. The chief advocate of this innermost planet with its gravity perturbing Mercury's orbit was Jean Leverrier of France, newly triumphant from his correct prediction of the position of Neptune, leading to its discovery in 1846 (see notes in chapter 4). He again tried to use Newtonian mechanics to explain the discrepancies in Mercury's behavior. He realized only a small body would be needed for the minor corrections, and that its orbit needed to be considerably inside Mercury's. Such a body would be incredibly hot, so he suggested it be named for the Roman god of the forge and metalsmiths. To find it, Leverrier suggested two possible means, each of which would yield unexpected rewards.

His first suggestion was to capture the tiny disk as it transited the sun, appearing as a tiny, fast moving sunspot. This meant of course that exacting sunspot records needed to be kept to catch the vagrant world. This search was fruitless, for no such body was needed once Einstein's relativity cleared up the problem with Mercury's precession. But the German amateur Schwabe was very dedicated in his pursuit of the planet, and he noted after many years of sunspot plotting that the numbers of spots seemed to come to a peak about every eleven years, the sunspot cycle.

A second possible way of spotting Vulcan was to catch the tiny disk during the darkness of a total solar eclipse. Very quickly the observer must search near the sun for any unplotted images, so accurate records of star positions behind the sun were essential. Later of course the shifting of those stellar images by the sun's gravity would confirm Einstein's relativity; still the search again turned up interesting new information. At several solar eclipses, previously undetected comets were spotted, some of them not to be seen again. They were sungrazers, comets coming so close to the sun that many are entirely destroyed. If they come from the direction of the sun, we may not spot them until after they round the sun at perihelion. If the sun destroys them, then we never see them, unless glimpsed at totality. The original discovery of such sungrazing comets came from the pursuit of Vulcan. In astronomy, as in life, the search itself is often more important than the original goal.

DEMONSTRATIONS AND OBSERVATIONS

Just How Dense are You?

I love to kid my students with the critical density. As we are chiefly made of water, with a density of one gram per cubic centimeter (or 1,000 kg/m^3, as Zeilik prefers), we are about 10^{30} times denser than the average density of the universe. We think of white dwarfs, at about a million g/cc, or neutron stars, at a trillion g/cc, as being dense, but compared to most of the universe, we are really pretty dense ourselves. I note in closing that some of us are denser than others, as will be revealed in the next test.

ANSWERS TO STUDY EXERCISES

1. In Newton's terms, since the acceleration is zero, there are no net forces acting upon you, and so the gravitational force is being balanced by the repulsive force on interaction between the molecules in your feet and the ground. You are at rest. In Einstein's terms, only one force is at work, the repulsive force between molecules of your feet and the ground. It is this force that is determining your particular path in spacetime, other than the natural, shortest freefall path. The earth is in the way of your free-falling motion.

2. In Newton's terms, the gravitational force on the orbiting path causes it to deviate constantly from a straight-line path, so follow its curved path orbit through space. In Einstein's terms, the orbiting body is in free-fall; no forces are acting, and so the body is following a natural path (straight line) determined by the body's velocity and the local curvature of spacetime.

3. In Newton's terms, the gravitational force of the sun on the planet causes it to constantly deviate from its straight-line path, so to follow an elliptical orbit around the sun. In Einstein's terms, the planet is in free-fall, with no forces acting upon it. Its path around the sun is created by the planet's own velocity and the gravitational distortion of space-time caused by the mass of the sun.

4. The average density of the universe is proportional to the total mass (and energy) in the universe. It is the total mass-energy that determines the overall curvature of space-time, that is, whether the universe is flat, open, or closed.

5. An observer in the Andromeda Galaxy would see essentially the same thing as one in Milky Way. The motions of the nearby galaxies would be different relative to the Andromeda galaxy (the Milky Way, for instance, would appear blue-shifted), but observations of distant galaxies would produce the same Hubble constant and law that we derived here in the Milky Way.

6. If a rocket is fired off the earth with less than escape velocity, it will travel upward more and more slowly until at some point its velocity will be zero. It will be at its maximum distance from the earth as it is then pulled back and will eventually fall to the earth. The history of the universe with a closed geometry is similar.

Consider two objects in such a universe. They will move apart starting at the Big Bang; then the rate at which they separate slows down, until it is zero. They will then be at their maximum distance apart. They will then come together, at first slowly after the turn around, then more and more quickly in the "Big Crunch".

7. Because light travels at a finite speed, the vast distances in the universe guarantee that our view of celestial objects is as they were in the past. The bigger the telescope, the farther out in space, and farther back in time we peer.

8. These are ALL situations with weightless conditions; a scale would read "zero" for the spacecraft, moon, earth, and Jupiter. This demonstrates that weightlessness indicates natural motion in spacetime.

9. The Hubble constant gives the rate at which the universe is expanding now. Assume that the rate is constant (it is NOT!). Then the inverse of the Hubble constant gives the time when the expansion began.

10. If there were no massive object nearby, then the natural motion (free fall) will be a straight line, exactly as called for in Newton's first law.

11. Einstein's $E = mc^2$ tells us that matter and energy are interchangeable, with a little matter capable of being turned into a lot of energy. Now what is conserved is the total amount of matter and energy combined.

ANSWERS TO PROBLEMS & ACTIVITIES

1. Hubble's law is $v = H \times D$; given an $H = 20$ km/sec/Mly, for $D = 500$ Mly, then $v = 20$ km/sec/Mly x 500 Mly = 10,000 km/sec for the observed recession velocity.

2. As $D = v/H$, then $D = 150,000$ km/sec / 20 km/sec/Mly, then $D = 7,500$ M ly.

3. If the Hubble constant were 30 km/sec/Mly, then the Hubble time would be only half the accepted value of 20 billion years, or just ten billion years old.

4. For a 100 kg person, collision with your anti-person would create 100 kg x $(3 \times 10^8)^2 = 9 \times 10^{18}$ watts per second. With 100 watt bulbs, this could light 9×10^{16} bulbs.

5. For 3 degrees, the Andromeda galaxy is 1/sin 3 deg = 19 times its diameter distant. It thus lies about 19x its own diameter distant; if its size is comparable to the Milky Way's, it thus lies 19x100,000 ly = 1.9 million ly distant, a little less than our current estimates.

6. At 200 Mly, it would be 100x smaller, or 1.8 arc minutes in diameter. It would be receding (assuming H = 20 km/sec/Mly) at 200 Mly x 20 km/sec/Mly = 4,000 km/sec.

CHAPTER 8

The Earth: An Evolving Planet

CHAPTER OUTLINE

Central Question: What are the basic features of the earth, and how have they changed since our planet's formation?

8.1   The Mass and Density of the Solid Earth

8.2   The Earth's Interior and Age

8.3   The Earth's Magnetic Field
       A. Magnetic fields and forces
       B. Variations and origin
       C. The magnetosphere
       D. Interaction of magnetic fields and plasmas
       E. Interaction with the solar wind

8.4   The Blanket of the Atmosphere
       A. Seeing
       B. Extinction and reddening
       C. Albedo
       D. The greenhouse effect
       E. Ozone layer
       F. Atmospheric circulation

8.5   The Evolution of the Crust
       A. Planetary evolution and energy
       B. Continental drift
       C. Volcanism and plate tectonics

8.6   Evolution of the Atmosphere and Oceans
       A. Origin and development of the oceans
       B. Evolution of the atmosphere
       C. Evolution of the earth's surface temperature

8.7   The Earth's Evolution: An Overview

Enrichment Focus 8.1: Radioactivity and the Dating of Rocks

CHAPTER OVERVIEW

Teacher's Notes

Our earth may not seem like an astronomical body, but it is! We live here, so we know it well. It serves as our model for the other terrestrial planets; what we understand about them we infer from analogies to the earth.

In the past two decades, space travel has provided a picture of the earth which we didn't have before: the earth from space. The earliest satellites probed the radiation belts around the earth. The Earth Resources Technology Satellites (ERTS) provided a detailed study of large areas of the earth. The Gemini series and Skylab provided many photographs of the earth which can be used to illustrate both atmospheric and surface features. The LANDSATS and French SPOT satellites have given us real time date on pollution, deforestation, and other ecological concerns. The new radar imaging technology lets us strip away the waves and sand dunes to map the oceans and deserts as never before; a lost city in southern Arabia was found by ancient trade routes in 1995.

Several lunar laser ranging experiments have been and are being carried out to possibly measure continental drift. Timing laser pulses reflected from mirror arrays left on the moon by Apollo astronauts allow changes in the earth-moon distance to be measured to a few millimeters.

Note that this chapter has a strong emphasis on the earth's magnetic field and its interactions with the solar wind. It would pay to do a few classroom demonstrations with magnets to remind students of the properties of magnetic fields. An electromagnet would be especially appropriate because it demonstrates that currents (moving electric charges) can generate magnetic fields and switch polarities.

Our home planet is the marvelous product of natural forces from both within and without. These forces are still at work, with the geological forces from within now more evident than the astronomical effects from beyond the earth. Increasingly, the surface of our planet has also been shaped by biological forces, even vastly transforming our atmosphere.

The study of our home planet has in recent years involved astronomers as well as more traditional geologists, meteorologists, paleontologists, and oceanographers. In the past decade evidence has been gathering that astronomical catastrophes have played a far larger role than most other scientists ever expected in our world's story.

8.1   The Mass and Density of the Solid Earth

The earth's average bulk density of 5,500 kg/m^3 is hard to visualize; a cubic meter is a large volume and not easily manipulated. I much prefer using grams per cubic centimeter with my classes. Most students have taken medicine (including vaccinations) in cc's, and a penny has a mass of about three grams. By definition, the density of pure liquid water at 4 degrees Celsius is one gram per cubic centimeter. Our bodies, made chiefly of water, are also about this average density. If your body is more dense than this, you sink; less, you float. As sea water is denser than fresh (all the dissolved salts adding to its density), it is easier to float in salt water, as your students know from the beach.

Most crustal rocks are in the range of 2.5-3 g/cc, but the bulk density of the whole earth is about twice that high. This means of course that what sits beneath the crust much be much denser than the crustal rocks are, leading to the concept of differentiation in the next section.

8.2   The Earth's Interior and Age

Differentiation is a vital concept. Stress to your students that all bodies hot enough at one time to melt underwent this process. As they cooled, they formed spheres, with their densest material in cores, and lighter materials on the outside. Thus if on Star Trek we are approaching a large world, it should be depicted as a sphere. In our solar system, bodies over 200 kilometers (150 miles) across are almost always fairly spherical, and asteroids and cometary nuclei under 100 miles wide generally irregular in shape--look at Viking photos of Mars' moonlets, Deimos and Phobos, and the new Galileo photos of asteroids Gaspra and Ida for good examples.

What heats the earth's interior? Zeilik mentions the gravitational heat of formation, and the continuing radioactive decay. The chief isotope acting today is probably Uranium 238, which has a half-life of 4.5 billion years. Thus the earth should have about half of this common form of Uranium left, not yet decayed to lead. But in the planet's past a variety of more radioactive isotopes, such as U-235 (weapons grade uranium), thorium, and even plutonium were also present and adding their fission energy to the core. Considering the decay of these isotopes, we see why the early history of terrestrial bodies (especially the Moon, Mercury, and Mars) should be more geologically active than the present heavily cratered terrain shows at first glance.

But it seems likely in the early days a lot of energy at the crust came from the kinetic energy of impacts; every time a comet or asteroid struck a body, great forces were released which created the lunar mare, split continents, and changed the entire history of life on our planet. Might it have been the impact of dense, metallic asteroids that slowly sank to our core that left behind the plumes that carry heat upward today from core to light the fires of Hawaii and cause Old Faithful to erupt?

8.3   The Earth's Magnetic Field

This is the one area in planetary science where we have the most problems facing us. The dynamo theory can not account for the strength and direction of the magnetic field of any other planet, nor explain why our own field switches polarity in an irregular fashion.

Relate the Van Allen Radiation Belts to the Enterprise's "force field", deflecting the rain of high energy particles from the sun toward the planet's poles, rather than letting them collide with our entire planet. Remind students that in LEO, such as the Space Shuttles, we lie well below the protective belts, but once we go up to geosynchronous orbit, or better yet head out to the Moon and other planets, our astronauts need to be flying a vehicle with a built in fallout shelter. The most intense solar radiation storms last only minutes, but a lead lined room aboard the Mars Mission craft seems certainly advisable.

If students have trouble relating to magnetic field's ability to deflect charged particles, remind them that a television screen image is formed by electrons being focused by magnets to the exact locations on the screen we desire. Thus the earth's van Allen radiation belts protect the equatorial regions, at the same time deflecting the charged particles of the solar wind toward the magnetic poles. There they strike the gases in our atmosphere, with each gas becoming ionized (that's why we call it the ionosphere!) and giving off its own set of emission lines and colors. The green in auroral displays comes from ionized oxygen, the red from the alpha line of hydrogen, and the blue from ionized nitrogen. Nature's own version of the three color TV display, with its varying colors and dancing patterns, can be more entertaining than anything on cable; ask students who have seen auroral displays and get their recollections.

8.4   The Blanket of the Atmosphere

Remind students that stars do NOT twinkle. Out in space, the stars shine just as steadily as any planet. But the slender thread of starlight is much more susceptible to atmospheric currents than the broader beam of light from the disk of the sun, moon, or a planet, so the star's light is bounced around much more and hence seems to twinkle. The lower the star is in the sky, the more atmospheric currents it must traverse, and the worse the image. As suggested in the observing projects section, have your students observe this effect telescopically.

Albedo is a term students often do not understand. Reinforce the concept by asking everyone to chose the person in your classroom wearing the outfit with the lowest albedo (almost black) and highest albedo (light to white). Then ask them to think of places on earth with high albedos (ice caps, snow fields, white sand beaches) and low (dark basaltic lavas, oil spills, dense tropical rain forest).

While we are concerned about the loss of ozone due to human CFCs, it is also worth noting that large volcanic eruptions can do great damage to the ozone layer, yet it has proven very resilient in the past. It is the CFCs ability to continue doing damage long after the original release of the gas which worries us, for this negates the layer's ability to rapidly replenish itself.

In introducing rudimentary meteorology, ask the students just what is wind. If they respond that it is the movement of air molecules, next ask why they move in a set direction. Hopefully someone will answer that the pattern carries air from high toward low pressure areas. Next ask them what common gas in our atmosphere can change phase to create the barometric changes and drive the weather machine.

Hopefully someone will come to the conclusion that water, changing into vapor or ice, is the chief agent in creating these changes. Remind them that the two principal components of your atmosphere, nitrogen and oxygen, remain gases at all temperatures and pressures naturally found on earth. In later chapters we will find that this is not often so. The net result is that only a minor component of our atmosphere is creating relatively mild barometric changes; our weather is not nearly as violent, on average, as that found on several other planets and satellites, as will be discussed later at Venus, Mars, Jupiter, Saturn, Neptune, Titan, and Triton.

8.5   The Evolution of the Crust

It is worth noting that the maps atop page 166 are merely representing the latest chapter in the continental soap opera. Long before there was Pangea, other continental pieces in very different arrangements covered the globe. As the differentiation process continues, it is possible that there may be more separation between continents and oceans now than ever before, although these continental margins have been very susceptible to the rise and fall of sea levels along the continental shelves with the ice ages during the last several million years.

In studying the cycling of the crust, again stress the stability of continental granite vs. the plasticity of heavier oceanic basalt. When two oceanic plates meet, both go down, with the heat of this subduction creating a volcanic chain along one side of the trench (Mariannas Islands). When a continental and oceanic plate meet, the basalt sinks into a trench, but the heat builds up the adjacent continent with a volcanic chain such as the Andes or our own Cascades. You can always tell which direction a continent is headed by the side on which the newest mountain chain lies; the Americas are heading west, but Australia is moving east, for instance. As Africa drifts north, it closes the Mediterranean, pinching if off at Gibraltar and drying it up at times.

If two continental plates meet, then there is no place to go but UP; the meeting of India and Asia in the last 30 million years produced the Himalayas. The earthquakes that destroy so much in Iran, Iraq, Armenia, and the Balkans are the result of this continental collision as well.

It is also worth noting that such mountains need DEEP roots; the colliding granite plates push down as well as up, making the earth's crust as thick as 70 kilometers under the Himalayas, while it is only about 10 kilometers thick under the sea floor.

While Zeilik stresses the role that radioactive decay of U 238 plays in driving the plate tectonic cycle, it is likely in the early solar system that other shorter lived but once common radioisotopes also played a role. Aluminum 26 is one with a half life of just a few million years, but it might have been produced in a nearly supernova and quite abundant in the early solar system. Its decay may account for many of the smaller moons of Saturn and Uranus appearing well rounded and differentiated, even though gravitational energy alone would not have caused them to melt. Uranium 235, the source of nuclear weapons, is much less common now than it once was; its half life is only abut 700 million years. As our planet is about 4.5 billion years old, it has already decayed through about 7 half lives, so less than one percent of it is left. By contrast, about half of the earth's original U 238 has not yet decayed. This, and the fact that of the really heavy radioactive elements it is the most common, is why U 238 gets so much consideration as the earth's internal energy source. And if half of it is still left, do not expect any substantial decline in volcanoes, earthquakes, and other tectonic activity for eons to come.

8.6 *The Evolution of the Atmosphere and Oceans*

Zeilik's point about outgassing is very important. Our oceans and atmosphere all issued out of the earth's interior via volcanic eruptions. These volcanic gases did NOT include oxygen, however. As noted, this has been the product of 3.5 billion years of photosynthesis.

Zeilik only briefly notes where the carbon dioxide went, and this is a vital element of the discussion. When photosynthesis occurs, the plant is seeking to make carbohydrates (carbon which has been hydrated or had water added to it). Glucose, for instance, is just six carbon atoms with six water molecules attached. If we hydrate the carbon, then the dioxide is free to escape.

The largest single form of stored carbon dioxide is not carbohydrates, but rather carbonate rocks. As soon as the earth's oceans formed, a large fraction of the atmospheric carbon dioxide went into aqueous solution. The earth's early oceans became seas of warm carbonated water. The discovery of similar carbonates in the Allen Hills Mars meteorite led to the discovery of apparent micro fossils in August 1996. But chemical precipitation and coral reef building organisms soon began extracting most of the carbonic acid solution and cementing it underground in the form of carbonate rocks; limestone underlies much of the U.S., and represents a tremendous amount of solidified carbon dioxide. Carbon dioxide is the chief gas in the atmospheres of Mars and Venus, and it would be here, too, if not for the deposition of the vast majority of it underground. But the cycle is not over, for when geothermal heating melts carbonate rocks, carbon dioxide is again liberated to gush out of volcanoes and back into the atmosphere. As the volcanoes of Mars died, the release of more carbon dioxide into its atmosphere stopped, and the planet chilled and died. As we have seen, it will be a long time before the earth's active interior dies, to doom our world as well.

As to the effects of burning fossil fuels, the geologic record reveals many naturally occurring changes in the climate probably much larger than anything we are now causing.

8.7 An Overview of the Earth's Evolution

To the three stages Zeilik discusses, others would add cratering, just after differentiation. We do not think much of this one today, since most records of our early crust and its impact scars are long erased. But for Mercury and the Moon, this stage is where their evolution ground to a halt. Still others would include our present tectonics in a broader category as "Slow Surface Evolution," including important changes wrought on the earth's surface of erosion by water, wind, and ice. Biologists would probably also like to see more emphasis on the living planet or the Gaia hypothesis. This notes how once life gains even a tiny foothold on the crust of a world, life evolves to not only fit ecological niches already in existence, but to transform the entire planet into a more habitable place for other forms of life to follow, much like the Genesis capsule from the third Star Trek movie.

ADDITIONAL RESOURCES

Transparencies 8.2, 8.3, 8.4, 8.6, 8.8, 8.10ab, 8.16z, 8.18, 8.19, 8.20, and F.6 will assist your class presentations.

The Planet Earth series on PBS had several hour long videos relating to our home planet. The fourth episode on comparative planetology is perhaps the best one to show as this chapter introduces the terrestrial planets. Sagan's Cosmos 104, "Heaven and Hell", does a good job of noting the uniqueness of the Earth, especially compared to its sister planet Venus; the title is very apt, and the graphics have not been dated much by new discoveries. Both the Planetary Society and the Astronomical Society of the Pacific sell both sets.

SUPPLEMENTARY ARTICLES

1. "Eyes on the Earth--Remote Sensing Via Satellites," Astronomy, August 1986, page 6.
2. "Planetary Bombardments," Astronomy, July 1988, page 20.
3. "Birth of Planet Earth," Astronomy, July 1989, page 24.
4. "Galileo Looks at Earth and Moon," Astronomy, April 1991, page 30.
5. "Demise of the Dinosaurs--A Mystery Solved?," Astronomy, July 1991, page 30.
6. "Asteroid Impact--The End of Civilization?," Astronomy, September 1991, page 50.
7. "Earth's Atmosphere--Terrestrial or Extraterrestrial?," Astronomy, Jan. 1992, page 38.
8. "A New Slant on Earth--Climate and Axial Tilt," Astronomy, July 1992, page 44.
9. "Volcanic Twilights," Astronomy, August 1992, page 36.
10. "The Cosmic Origins of Life," Astronomy, November 1992, page 28.
11. "Death from the Sky (Tunguska)," Astronomy, December 1993, page 38.
12. "Bands, Glows, and Curtains (Aurora)," Astronomy, April 1995, page 76.
13. "A Friend for Life (Habitable Planets)," Astronomy, June 1995, page 46.
14. "Target Earth," Astronomy, October 1995, page 34.
15. "The Day the Dinosaurs Died," Astronomy, April 1996, page 42.

Also check out articles in Scientific American, National Geographic, Earth, Discover, and Science News for more new material on our expanding knowledge of our home planet. It is hard for me to make it through this chapter--there is so much material I want to discuss. Encourage your students to look through these magazines for new developments--the entire class can learn much. Science is an exploration that needs to capture the imaginations of the students for our educational system to perform at "world class standards!"

ADDITIONAL DISCUSSION TOPICS

Cosmic Catastrophes

The idea that major impacts have played a great role in shaping the earth is becoming more prevalent. The recent close encounters with Comet Hyakutake in March 1996 and Asteroid 1996 JA in May 1996 certainly has made people aware that there are sizable asteroids and cometary nuclei that have struck the earth in the past. The announcement in late 1992 that the Yucatan impact dated exactly 63 million years ago has done much to convince most geologists that it played the largest role in the extinction of the dinosaurs. Much more controversial was the December 1992 speculation that an even larger impact at the end of the Permian almost wiped out all life of the earth, and even split the crust off southern South America to begin the division of Panagea into the northern Laurasia and southern Gondwanaland supercontinents. There is an interesting article on this closest of all calls with global extinction in the November 1993 issue of Earth. Some geologists argue the vast amount of dust even created an intensely cold but very short ice age then, and that the present ice age, lasting millions of years, is in fact the only prolonged cold period in the earth's entire history. This is very controversial at this time.

In the final chapter of the book we will speculate on contact with extraterrestrial life. Consider the role that the frequency of such major impacts could play on biological evolution. In a planetary system with more debris in turbulent orbits, this could mean more large impacts could wipe all life, or at least any advanced life, off an otherwise earthlike planet every few hundred million years, never allowing anything as advanced as ourselves to evolve to date. Perhaps that might be one reason why ET hasn't come calling.

Lunar Eclipses as Atmospheric Probes

Zeilik mentions the role that atmospheric dust plays in scattering the shorter (blue) wavelengths and transmitting only longer (redder) waves near sunrise or sunset. In a total lunar eclipse, the only light to reach the moon comes from sunlight refracted by our atmosphere around the limb of the earth to the moon. If there is a lot of extra dust, almost all of the sun light will be blocked, and the eclipsed moon may be practically invisible. The September 1996 issue of Sky and Telescope has an in-depth article on how the atmosphere creates the effects observed in lunar eclipses.

For instance, on December 9, 1992, the rising moon was invisible to the naked eye during totality to many observers; it was much darker than its normal reddish appearance.

The reason for this extra dust is the 1991 eruption of Mount Pinatubo in the Philippines. This blast threw much debris into our stratosphere, much as the impact of a comet or asteroid could have done. Because this dust is taking so long to settle, it is likely winters will be substantially colder for the next several years, since the dust is reflecting much of the sun's energy back into space before it can ever heat up the ground.

Also on a more fundamental level, Aristotle used lunar eclipses to first show the earth was a sphere considerably larger than the moon. The fuzzy shadow of the leading edge of the earth testifies to our envelope of air, refracting and scattering the sunlight. While not as spectacular as a solar eclipse, lunar eclipses are a great class activity if the right one comes along. They are "laid back events," easily seen with the naked eye, binoculars, or low power telescopes. Use them as recruitment tools for getting students interested in taking your course as well.

Magnetic Polarity Reversals

A problem with the conventional dynamo model of planetary magnetism is that we are very uncertain as to how magnetic polarity reversals fit into the picture. We find that 800,000 years ago, the north pole of your compass would have pointed south, not north. But a million years ago, the polarity was the same as at present. These polarity reversals are revealed as we study igneous rocks. Any time an iron bearing rock crystallizes, the tiny iron crystals align themselves with the earth's magnetic field at that moment. If we then are to carefully study the rocks in place, they are a fossil of the magnetic field of the earth when they cooled.

How long these reversals take is unknown, for the rock record can not be dated that precisely. While the field seems to switch polarity every few hundred thousand years, so far no pattern for these reversals has emerged. Our own present 700,000 year "normal" polarity is definitely on the long side, and perhaps on thin ice. In this century considerable drift has occurred in the surface locations of both N and S magnetic poles; a switch in polarity could happen at any time. I doubt it will be dangerous--our ancestors lived through many already, and no extinction has been tied to any of the magnetic reversals. The July 1996 announcement that the earth's solid inner core is spinning slightly faster than the rest of the globe may hold a key for greater understanding of these reversals.

DEMONSTRATIONS

Continents vs. Oceans

If possible, get a hand-sized piece of granite (any railroad track might use this light colored igneous rock for road bed) and basalt (most any shipyard might use this dark igneous rock for ballast). Have the students pass around about equal sized pieces of each, comparing the weights and densities of each. Most students can tell the lighter colored granite is also lighter in weight; remind the students this is the bedrock of the continents, which float atop the heavier ocean basins of basalt. Except for this difference in density, the continents would long ago have slumped beneath the ocean waves, as the dense, basaltic Hawaiian Islands are doing in mid-Pacific now.

In fact, when mid-ocean volcanoes like Hawaii do build islands, those islands of heavy basalt will stay up only as long as the plume keeps pouring hot lava up and out. Once the drifting ocean plate takes the volcano away from its source of fresh igneous rock, the weight of the basalt starts pulling it back down, beneath the ocean. Look at a map of the Pacific Ocean floor. Note that as one moves northwest from the Big Island, the Hawaiian Islands become smaller and lower as they slump down, until at Midway, only the still growing coral atoll keeps it afloat. But the Hawaiian plume produced great mountains long before the 70 million year old basalt created Midway.

The Emperor Seamounts extend northward into the Aleutian Trench, and the ones now going down the trenches date back about 200 million years, probably the oldest oceanic crust yet found on the sea floor. The ocean crust is thin, and constantly recycling, with new basalt oozing out at the rifts, and old crust being tucked under the plates at the trenches. By contrast the lighter granite floats atop most of the activity, and thus is on the average much older than any seafloor basalts; in the central United States much bedrock exposed at the surface is over a billion years old, and some Canadian Shield material goes back 3.5 billion years.

A Really Dense Globe

Don't let the geography teacher throw away that globe just because it still shows the Soviet Union and Yugoslavia as countries. Let's use it to make a realistic model of the earth's interior, with the density to scale. To do this, you will need some scrap iron in small pieces, some plaster of paris, and a calculator. Cut the globe in half (most separate at the equator, anyway), and measure its diameter in centimeters. Halve this to get our globe's radius, then stop up the south pole hole with tape on the inside. Calculate the volume of your hemisphere by calculating $2/3 \times 3.14 \times radius^3$ (half the volume of the entire sphere); for a typical 12" diameter globe, this will be about 7,000 cc. As our planet has a bulk density of 5.5 grams per cubic centimeter, your hemisphere should then weigh about 39 kilograms, or 86 pounds! In filling your globe, you will obviously need a LOT of scrap iron, and then fill in the voids with plaster of paris. Such a model is a little heavy to pass around, but it might be a good idea to have students come past a table and try moving it. Have the pieces of crustal granite and basalt handy to let them realize just how much less dense these crustal rocks are than the bulk density of our world. Then stress that in differentiation most of the denser iron and nickel sank to the core, leaving just the lighter silicates near the top of the mantle and in the crust. That any of the heavy elements such as iron, and especially gold, are found in crustal rocks now is usually evidence of the mixing forces of the volcanoes bringing such material up from the core. In the case of Sudbury basin, however, the nickel probably was brought in fairly recently on an asteroid, and thus has not had a chance to sink down into the core yet.

You might complete your earth cutaway by painting the interior cross section according to the figure in the text. If you wanted a globe to pass around, a globe six inches in diameter would yield a hemisphere weighing only about 10 kg (22 pounds), if made to the earth's bulk density. The earth is a lot denser than any bowling ball!

Modeling Earth's Magnetic Field

A bar magnet about six inches long can be implanted in the other hemisphere of that 12 inch globe you dismembered. Attach it to the back of a poster board disk 12 inches across to fit in the cross section of the globe. Remember as Zeilik notes that the magnet needs to be tilted 12 degrees off the N-S axis of your globe. Sprinkle iron filings on top of the poster board to form the magnetic field lines, or use small compasses to trace out the magnetic field lines.

We generally think of our magnetic poles as being on the earth's surface, much like the geographic poles. Yet our core, the source of our dynamo, lies far beneath our crust. To illustrate to your students where our magnetism comes from, use a large compass, tilted vertically (a magnetic dip needle) and note that the compass points deep down toward the top of the core, not at the earth center, but nowhere near its surface, either.

Observing Twinkling

As twinkling increases with poor seeing, and the image quality of telescopic objects is going to be very poor under such conditions, just what can you show your students on a crisp, cold, clear winter's night when the sky is black as coal, but the image of Jupiter will hardly even show the belts and zones? Use the atmospheric turbulence, often greatest just after a cold front has passed, to give the students a light show they will treasure.

Use fairly high power (100x or so) and point the scope at a bright star (Sirius works very well) low in the atmosphere, one notably twinkling even with the naked eye. Let them see the flashes of spectral color which the refracting and dispersing layers of the atmosphere create. Now through the star slightly out of focus, and note the colors again; they swirl across the star's out of focus image like a kaleidoscope. Now move your hand (take the glove off for maximum effect) in front of the scope objective, so that the currents of warm air flow off your hand and further distort the star's image. It is a lot of fun, and certainly an entertaining way of making the best of a bad situation.

Observing Reddening and the Green Flash

A nice sequence for a student project is to have the student photograph the setting sun with a moderate telephoto (135-400 mm). Get as clear a western horizon as possible; the beach or some other waterfront would be ideal. Start when the sun is about one degree (two solar disk diameters above the horizon, and take photos about every 10 seconds or so, showing the earth's rotation as the sun sinks, growing redder as it descends. When projecting the sequence as slides, have the students note the sun moves its own diameter downward every two minutes, as the earth is rotating at the rate of fifteen degrees per hour, or one degree every four minutes.

You will be able to use your camera's light meter to control the exposure, and the sun's image is so dimmed by the atmospheric absorption that you can look at the sun briefly through the camera viewfinder without damaging the eye. It is worth noting that the sun has really already set by the time you see the bottom edge of the disk touch your local horizon. The atmospheric refraction near the horizon is almost a full degree, larger than the sun's half degree diameter. We say we spend half the year in day, and half in night, but when you consider the effect of atmospheric refraction, we really get several extra minutes of sunlight each day.

You may even see large sunspots on the sun's red disk on your photos, if the seeing is good enough. Ancient Chinese astronomers observed these at sunrise and sunset, and referred to the dark markings as "birds on the sun." You may also note the atmosphere is layered, with distinct levels notable in photos.

The most dramatic effect is the famed "Green Flash," which I have been lucky enough to observe on three occasions. The blue and violet light are scattered by the atmosphere, but often as the atmosphere acts like a prism, it will direct the last rays of the setting disk back to us as a glorious emerald green flash, lasting only a second or so as the disk's tip vanishes behind the horizon. I have both photographed and videotaped this ephemeral effect, and the slide film revealed the green color better, but not as vividly as did the eye. This again is best been as the sun sets over a large body of water and the atmosphere is calm; if the stratification of the layers is evident, be ready--you may spot the green flash today.

The green flash may be better seen at altitude. On the evening of the total solar eclipse of July 11, 1991, some friends and I watched the sunset from our eleventh floor balcony at a resort at Mazatlan, Mexico, on the Pacific. We saw the green flash very well, while observers on the beach missed it entirely. Perhaps the effect would be even more evident in an airplane. So the next time you are flying at sunset, glance briefly at the setting sun. Do so carefully, for the sun will be brighter from higher altitude (less atmosphere to absorb light and redden the image). Astronauts often describe the beauty of the frequent sunsets and sunrises (every 45 minutes from low earth orbit) as seen in space.

ANSWERS TO STUDY EXERCISES

1. The acceleration of a falling body is proportional to the mass of the earth: $g = GM/R^2$, so you would have to measure your acceleration (based on the time of fall), the earth's radius, and know G to get M, the earth's mass.

2. The earth's core is mostly molten iron and nickel and their alloys, while the crust consists of chiefly silicate rocks, mainly basalt for the ocean basins and granite for the continental bedrock, which are combinations of silicon, oxygen, iron, aluminum, and magnesium. The core is metallic, while the mantle and crust are rocky.

3. Uncertainties in the earth's age arise from assumptions made in the estimates. These is an uncertainty in the time between the formation of the earth itself and the solidification of its crust. One billion years may be a reasonable guess but may be off by as much as 10-20 percent. Also, it is assumed that fundamental physical constants have not changed as the universe ages. If such constants have changed, the rates of radioactive decay would have been different that they are today. Knowing the half-lives for radioactive decay is basic to estimating the earth's age--these are well established.

4. The average density of the material in the earth's crust is about half that for the entire earth. So the material in the core must be even denser still to result in the overall average.

5. The earth's atmosphere absorbs radiation at particular wavelengths, especially those in the very short wavelengths, ultraviolet, and infrared. In addition, molecules in the earth's atmosphere are more likely to scatter light of shorter wavelengths. This results in the reddening of direct rays of light, while the blue light that is scattered accounts for the color of the daytime sky.

6. Since water vapor (and carbon dioxide) absorbs infrared radiation but is transparent to visible light, an increase in the amount of water vapor in the atmosphere would result in less infrared radiation escaping into space, with more being trapped close to the surface (the Greenhouse Effect). This trapping would increase the temperature of both surface and atmosphere.

7. In outgassing, volcanoes release large amounts of water vapor, carbon dioxide, nitrogen, methane, ammonia, and hydrogen sulfide into the atmosphere. This output may have been the primary source of our atmospheric nitrogen, today 80 % of the air we breathe. Carbon dioxide was taken out of the atmosphere by photosynthesis and by being dissolved in the oceans, then laid down beneath them in carbonate rocks like limestone. Since both the carbon dioxide and water concentrations affect the earth's temperature and thus plant growth, the balance between the addition of these constituents to the atmosphere and their removal by other agents is a sensitive one, and has changed over time and is still changing.

8. The earth's initial atmosphere was mostly hydrogen and helium. It escaped into space. The next atmosphere degassed from the interior. It has been transformed by ultraviolet light, interaction with the oceans, and the proliferation of life.

9. The oceans absorb carbon dioxide and other gases from the atmosphere, trap much solar radiation and slowly release it back into the atmosphere, and are the reservoir for water in the atmospheric water cycle.

10. Most of the earth's water probably came from the interior during volcanic outgassing. Water vapor is released from the interior when lava erupts. Some also came in aboard comets and asteroids, but the amounts are in debate.

11. The solar wind is a plasma, containing charged particles (mostly protons and electrons). The earth's magnetic field interacts with the moving charged particles, causing them to spiral around the magnetic field lines, and so trapping them.

12. The thin oceanic basin crust is the youngest, being constantly recycled between rifts and trenches. The very newest materials lie right at the midocean ridges.

ANSWERS TO PROBLEMS AND ACTIVITIES

1. As one half-life has passed, about half of the original U 238 remains today. After another half-life has passed, only half of that half, or a quarter will still be radioactive.

2. $g = GM/R^2$, so 9.8 m/s^2 = 6.67 x 10^{-11} x M / $(6.400$ x 10^6 m$)^2$; solving for M_{earth}, therefore we get $M = (9.8) (4.1$ x $10^{13}) / 6.67$ x $10^{-11} = 6.0$ x 10^{24} kg.

3. If rotation is so vital to the dynamo (which Mercury's field calls into question), the slowing earth will generate a considerably weaker magnetic field.

4. If the radius and volume were unchanged, increasing the mass by four times would likewise increase the density four times as well. As the escape speed is also proportional to the mass, the escape speed would be four times greater, perhaps high enough to retain lighter gases and allow the earth to grow into a jovian planet.

5. To find the volume fractions of the mantle and core, let X be the volume fraction of the core, so the mantle volume is then (1-X). The total density times the earth's unit volume is then the sum of the core and mantle density fractions, or 5500 x 1 = 10,000 X + 2500 (1-X); x = .40   Thus if the core represents 40% of our planets volume, the radius fraction is proportional to the cube root of the volume fraction, so R = $(.40)^{1/3}$ = .73; the core extends 73 % of the way out from the center to the crust. This is higher than most estimates, which place the core density closer to 14 g/cc, or 14000 kg/m^3, making the core smaller accordingly.

CHAPTER 9

The Moon and Mercury: Dead Worlds

CHAPTER OUTLINE

Central Question: What processes have driven the short evolution of the small, airless worlds of the moon and Mercury?

9.1   The Moon's Orbit, Rotation, Size, and Mass
        A. The once-faced moon and lunar day
        B. The moon's physical properties
        C. Tides and the moon
        D. History of the moon's orbit

9.2   The Moon's Surface Environment

9.3   The Moon's Surface: Pre-Apollo
        A. Maria and basins
        B. Craters
        C. Cratering

9.4   Apollo Mission Results
        A. The lunar surface
        B. The moon's interior
        C. Lunar history and evolution

9.5   The Origin of the Moon
        A. The fission model
        B. The binary accretion model
        C. The capture model
        D. The giant impact model

9.6   Mercury: General Characteristics
        A. Orbit and rotation
        B. Size, mass, and density
        C. Surface environment
        D. Mercury's surface
        E. Magnetic field

9.7   The Evolution of the Moon and Mercury Compared

Enrichment Focus 9.1: Momentum and Angular Momentum

CHAPTER OVERVIEW

Teacher's Notes

This chapter compares the moon and Mercury as similar objects such that our extensive knowledge about the moon might be extended to the planet Mercury. The moon and Mercury are very similar in appearance and basic characteristics. One way to introduce this chapter is to show pictures of Mercury and the moon, and try to sort them out. Unless you (or the student) are very familiar with the moon's surface, you probably won't be able to!

The high interest in the Apollo flights makes the products of those missions useful teaching tools. Back up discussion of recent findings with slides or videos from the Apollo flights. The videos are available from NASA, the Planetary Society, and Astronomical Society of the Pacific. In addition, NASA has developed educational materials, aimed at the secondary school level, which use actual samples of the lunar material. See Additional Resources section below for more details.

If the astronomy course includes observations, the moon offers many possibilities. In addition to observing the sketching features, or identification exercises, photography can be introduced as a data collection tool. Lunar photographs allow features to be measured; for observation, features near the terminator, or boundary between lighted and unlighted parts of the surface, are most clearly seen.

Chapter 1 dealt with eclipses in a bare-bones way. We recommend that the gory details of eclipses, especially solar ones, be put off until this chapter, if you wish to cover them in your course. Solar and lunar eclipses are perhaps the most spectacular and most popular natural events which are easily observed. Solar and lunar eclipses differ in the circumstances of their occurrence, duration, and visibility. Solar eclipses are visible only from places on the earth covered the moon's shadow. Lunar eclipses are visible, and appear the same, to all observers on the night side of earth while the eclipse is in progress; better than half of the planet can view a lunar eclipse at least in part. Lunar eclipses may last over three hours. During a total lunar eclipse the full moon usually does not disappear entirely (the eclipse of December 7, 1992 was exceptional--see chapter 1 notes). Instead the full moon appears a deep red or coppery hue due to the refraction and scattering of sunlight in the earth's atmosphere.

Eclipses occur more often than most people realize. There are a minimum of four eclipses in a calendar year (two solar and two lunar), and a maximum of seven (four solar and three lunar or five solar and two lunar).Ancient methods of predicting eclipses can be introduced when discussing details of the motions of the moon. As noted in chapters 1 and 2, ancient civilizations recognized cycles of changes in the heavens, including eclipses. A simple and accurate method of eclipse prediction was discovered by the Chaldeans, who found that solar and lunar eclipses recurred on a cycle of eighteen years and eleven days.

The explanation of the Chaldean cycles is itself an interesting study of coincidences in the moon's motion. For an eclipse to occur the sun, moon, and earth must lie along a line, the line, the line of nodes. The moon is at a mode when it passes though the plane of the ecliptic (the path along which eclipses may occur). Because the line of nodes slowly moves with respect to the stars, the moon returns to a mode every 27.212218 days, called the Draconic month.

The time between successive new moons, the synodic month, is 29.530588 days. For eclipses to occur on a cycle, the cyclic period must be a whole number of Synodic month and Draconic months. It turns out that 223 lunations or Synodic months are equal to 242 Draconic months to within 51 minutes. In addition, 223 lunations equal, to within five hours, 239 anomalistic revolutions, the time between successive perigees. And so, through a remarkable accident, the moon returns to essentially the same place in its orbit every 18 years 11.3 days. So to predict eclipses, you need only to record all eclipses that occur over one cycle, then add 18 years 11.3 days to the date of each eclipse. Remember that the third of a day means that the eclipse will occur at about the same latitude, but a third of the world west of where the one did the previous Saros cycle. For instance, the longest eclipse of the twentieth century took place on June 30, 1973 in the middle of the Atlantic. The next longest one took place on July 11, 1991 in the Pacific and Mexico.

9.1   The Moon's Orbit, Rotation, Size, and Mass

The heavily cratered surfaces of the Moon and Mercury show us that planetary evolution on these worlds was short-lived, compared to the three larger terrestrial planets. Being smaller bodies, they cooled far faster than the larger worlds did, and their periods of volcanic activity have long since passed. Yet part of their value to us is the fact that they represent an early stage of planetary evolution, frozen in time like a fossil. While 97 % of the earth's crustal rocks have been eroded, melted, or otherwise changed in the last three billion years, on the moon and Mercury, probably 97 % of their rocks have reminded unchanged.

Ask your students who has the greater gravitational pull on earth, the sun or the moon. Many will probably say the moon. Correct them and note that we are all part of the solar system--it is the sun's great mass and gravity that is calling the shots. Then ask them who has the greater tidal influence; any fisherman will correctly name the moon. Go on to Zeilik's explanation of the tides as differences in gravitational pull. I like to hold up a tennis ball and explain gravity is a one-way street-- the sun is pulling us toward it. But tides are stresses, and I pull on the tennis ball from both sides to approximate tidal stress.

Consider how lucky we are to have the moon where it is now. It is at just the right distance to give us total eclipses at perigee, and annular eclipses at apogee. As it continues to pull away from us, in a few hundred million years it will be too small to give us total eclipses at all. The closer moon of the past would produce more and longer lasting totalities.

9.2   The Moon's Surface Environment

Stress that both surface gravity and temperature are involved in determining who can keep an atmosphere. While the Moon's surface gravity is only 1/6th the earth's, the surface gravities of Saturn's moon Titan and Neptune's moon Triton are even less. Yet because these are far colder worlds, with their surfaces receiving little solar heating, both have sizable atmospheres of nitrogen; in the case of Titan, it is even thinner than the earth's. If our moon were orbiting Saturn, Uranus, or Neptune, it could keep an atmosphere, too. In fact, there are transient gases and heavy inert gases that do linger around the moon, as discussed in the article on page 36 of the November 1993 issue of Astronomy. Even so, the moon is so close to a vacuum that it will be an ideal site for 21st century astronomical observatories in all wavelengths, as discussed in chapter 6.

74

9.3   The Moon's Surface: Pre-Apollo

Hold up those pieces of granite and basalt you used in Chapter 8 again. While the anorthosite of the lunar highlands is not identical to granite, they are chemically similar and are of about the same color and density. The bright highlands are made of light colored, light weight, older "continental" anorthosite, while the darker mare are flooded with heavier, darker, more recent basaltic flows. Note that basalt is a dominant surface rock on ALL the terrestrial bodies, as the Russian landers showed on Venus.

It is worth noting that the craters formed are typically about 20 times as large as the size of the impacting body. Since most strike with such kinetic energy that they blow up, the craters are usually round in appearance. But in several cases on both moon and earth, almost grazing impacts at relatively low speed have produced elongated craters. One in Argentina even appears to have the impacting body skipping across the pampas. The Alpine Valley and the elongated crater Schiller on the moon may be other examples of such grazing impacts.

Remind the students how important crater counts are in determining the ages of surfaces, anywhere in the solar system that we do find a solid surface. Also remind them the rate of cratering has dropped immensely since the early days, as more and more comets and asteroids were swept up by the gravities of the planets, or swallowed up by the sun itself, or expelled a la Voyagers to the outer reaches of the solar system and beyond.

Note also the role the rays around the craters play in determining their age. A recent impact will still have plenty of bright, glassy beads scattered on the surface around it, and stand out at full moon, such as Tycho and Copernicus do. But with time, the constant fall of dark meteoric dust will cover the ray patterns and hide them beneath the regiloth.

9.4 Apollo Mission Results

If you have a moon globe with both sides accurately depicted, then quickly spin it around, and ask students if they are looking at earthside or farside. Those familiar with looking at the full moon will quickly note the mare are selectively found on our side, for about 50% of earthside covered with mare, and only about 10 % of farside so altered; the flows on farside are also much more localized, in sites like Mare Moscovine and Mare Orientale, compared to the huge flows on the side we see constantly. When I was growing up, I remember people in the pre-Sputnik era talking about the dark side of the moon; ask your students which side of the moon is really the dark side. Then have them look at Figure 9.17 and ask them why, based on the Apollo discoveries of the moon's interior, this happened. They should be able to note that the displacement of the molten lunar core toward us brought the hot, molten material closer to earthside, and made the crust above it thinner on earth side. Equally large impacts on both hemispheres (Korolev and Mare Crisium, for instance) would produce equally large craters, but only on earthside would the impact crack the thinner crust deeply enough to cause the large scale flooding of the mare basins. On page 185, Zeilik notes that a few huge chunks smashed the crust to form the mare--mostly on the side that faces us. Statistically the large impacts were equally likely on both hemispheres, and did occur on both; it was the tidal displacement of the lunar care that made the mare. The crater Korolev on far side is as large as Mare Crisium, but only on our side was the thinner crust breached by the impact, with the basin subsequently flooded with lava. The thicker crust of far side kept Korolev just one more huge crater.

Stress again to the students how quickly the moon died. The moon is a much smaller body than the earth, so it cooled so fast that volcanism on the moon basically stopped about 3.2 billion years ago, according to the ages of the youngest Apollo mare basalts.

Ask the students if they have ever heard of a "leap second". About every year or eighteen months, to compensate largely for the tidal slowing of our rotation by the moon, the National Bureau of Standards must add an extra second to the standard clock, so that our atomic clocks will be in synch with astronomical observations based on the earth's slowing rotation. Listen carefully to National Bureau of Standards Station WWV (5, 10, and 15 MHZ) and you can hear 61 ticks on the clock on such occasions.

9.5   The Origin of the Moon

The original reason the fission model was popular is the notable similarity between the earth's surface density (granite and basalt are both around 3 g/cc) and the bulk density of the moon (3.3 g/cc). The original proposal even had the Pacific Ocean basin as the scar left when the moon spun off. Of course, continental drift has now shown us this is just a temporary situation, and the similarity in bulk density is negated by the large differences in water content and other volatiles between the moon rocks and earth crustal rocks.

The binary accretion model is probably correct for most planetary satellites. They did form in orbit around their planet's equators; if you can, show your students the remarkable alignment of the four Galilean moons parallel with Jupiter's equator. But our moon does NOT orbit above our equator, but instead follows the path of the planets, the ecliptic, through the sky with a variation of only 5 degrees (less than that of Mercury or Pluto, for instance). This apparent relationship to the ecliptic is one reason the capture model was initially popular, for the moon would have already been orbiting in the ecliptic plane and been made in a different part of the solar system.

The different site of formation could have accounted for the chemical differences between earth and moon rocks. But the statistics for such a capture are astronomical. It would be much more likely that the two bodies would actually collide, which leads us to the giant impact model.

In such an impact, note the great heat of collision drove the lightest, most volatile materials far out into space, so the water vapor and other volatiles escaped and were not part of the condensing moon. The earth's gravity brought most the heavier debris back to it, to become part of our own core. Only the middle-weight materials, of density similar to our crustal rocks, would have been left in orbit to form the moon. As the impacting body was probably moving in the ecliptic plane, the material thrown out in the impact would probably have condensed out in the same plane, accounting for the moon's present orbit. As Zeilik points out, the impact was a glancing blow, so the collision may have resulted in our own planet being tilted, you guessed it, by 23.5 degrees from the ecliptic. Not only does the giant impact model (Big Whack Theory to some) account for the moon, it also accounts for the earth's seasons. In the "To Boldly Go" video on the Jovian planets, it is suggested a similar collision accounts for the 98 degree axial tilt of Uranus. Other examples of the transformations wrought by such impacts include the huge craters on Saturn's moons Mimas and Tethys; in the later case, the crack stretches all the way around a moon half as large as ours. In the most extreme case, Uranus' moon Miranda apparently was broken up into a dozen or so huge pieces, which fell back together at random, as its strange surface indicates.

9.6 Mercury: General Characteristics

There is new evidence that Mercury's surface environment may not be as hostile as Zeilik describes. Arecibo radar data late in 1991 indicated that at Mercury's north pole, there may be a permanent polar cap of water ice. While the tilt of Mercury's orbit has not yet let us confirm this, or check out the south pole, it is an exciting discovery with great implications to planetary exploration in the next century. As Mercury has no axial tilt, both poles keep the sun on their horizon. The sun never climbs high enough to melt ices hiding in the shadows. While at Mercury's equator the day versus night temperatures might vary by over a thousand degrees Fahrenheit, at the polar regions this variation would be considerably less. Establishing a human colony at both of Mercury's poles might be far easier, if abundant water ice were present, than building a colony on the moon. Mercury is far more accessible from earth orbit via spacecraft than is Mars .

In a parallel to the exploration of the western hemisphere, the Spanish and other Europeans did not get excited until abundant gold and silver were discovered; as Mercury is the densest planet in terms of chemical composition, tell your students "there's gold in them thar hills!" The sun's radiation would also provide ample energy for solar refining of the riches of this gold mine of the solar system.

In your discussion of Mercury's scarps, point out that most substances, such as Mercury's iron core, shrink when they cool. Like our moon, Mercury cooled along ago, and its core shrank. If fact, there is only one common substance that expands upon freezing. What is it? Water, of course. On the outer moons, such as Uranus' Titania, we will see great cracks in their crusts from where the ice expanded as these low density worlds froze.

No other world so demolishes our dynamo as does Mercury with its magnetic field. If density and an iron core are so important, why doesn't far more earthlike Venus possess a field as well? If rotation were vital, Mars should have a stronger field too. Explain Mercury's magnetism and get your ticket to Stockholm. Zeilik's comparison of Mercury's field to a permanent magnet is probably correct, but I certainly you would like to see a Mercury orbiter placed on NASA's agenda in the next two decades. Not only could it look for variations in Mercury's magnetic field, but it could check out those polar caps and metal deposits.

9.7   The Evolution of the Moon and Mercury Compared

In the early history of the solar system, the solar heating drove the lighter, more volatile elements farther out, accounting for the lower densities of the Jovian worlds in the next chapter. But this also helps explain why Mercury, closest to the sun, was made of only the densest elements. The moon, though only slightly smaller than Mercury, formed farther out, and like Mars, is lower in density as well.

ADDITIONAL RESOURCES

Transparencies 9.1, 9.3, 9.6ab, 9.17, 9.23, 9.24, and 9.29 and slides 9.20a and b will assist your class presentations.

The NASA video, "The Moon--An Emerging Planet" is a short video, available from your closest NASA film library. It is only about 16 minutes long, and summarizes well the discoveries about the moon's early history thanks to Apollo. Get the address of the closest one by calling NASA Houston at (713) 483-3111.

You might want to check with NASA's Houston Lunar Receiving Laboratory about the possibility of taking a short course and becoming a registered moonrock handler. You can then check out on two week loan America's crown jewels. They also have meteorite specimens for checkout as well, and your students will find these look even more interesting than the admitted rather bland moon rocks; you can check out both types at the same time. It is a BIG security hassle, but the effort is probably worth it, particularly if you have a special event, such as our planetarium opening here at PJC, that you want to publicize. Contact Mr. Boyd Mounce at (713) 483-8623, or write him at AP-4, JSFC, Houston TX 77058.

In Cosmos 104, "Heaven and Hell", Sagan discusses the possibility that a comet struck the moon in the twelfth century AD, as witnessed by the Canterbury monks. He does a good job of illustrating the evolution of the moon with great animation in this video.

SUPPLEMENTARY ARTICLES
1. "Sculpting the Moon," Astronomy, February 1987, page 82.
2. "Looking for Lunar Fractures," Astronomy, February 1988, page 56.
3. "Mercury's Heart of Iron," Astronomy, November 1988, page 22.
4. "Eight Wonders of the Moon," Astronomy, March 1989, page 66.
5. "Apollo Memories," Astronomy, July 1989, page 22.
6. "How Apollo Changed the Moon," Astronomy, July 1989, page 40.
7. "Back to the Moon as Colonists," Astronomy, July 1989, page 48.
8. "Finding Lunar Volcanoes," Astronomy, December 1990, page 62.
9. "What would the Earth be like without the Moon?," Astronomy, February 1991, page 48.
10. "The Joy of Moongazing," Astronomy, March 1991, page 84.
11. "Galileo Look at Earth and Moon," Astronomy, March 1991, page 30.
12. "Mysteries of the Moon," Astronomy, December 1991, page 50.
13. "Search for the Lost Lakes of the Moon," Astronomy, March 1993, page 26.
14. "The Moon's Atmosphere," Astronomy, November 1993, page 36.
15. "Clementine: Mission to the Moon," Astronomy, February 1994, page 34.
16. "Return to the Moon," Astronomy, May 1994, page 32. Astronom
17. "Apollo's Gift: The Moon," Astronomy, July 1994, page 40.
18. "A Swirl of Moondust," Astronomy, October 1994, page 28.
19. "Fire Fountains of the Moon," Astronomy, March 1995, page 34.
20. "Giant Holes of the Moon," Astronomy, August 1996, page 50.

ADDITIONAL DISCUSSION TOPICS
Lunar Libration

The moon does not keep EXACTLY the same face always toward us, as any telescopic observer of the moon will soon note. At some full moons, craters near the limb will not be seen, while on the opposite limb craters not seen before will have tilted into view.

The moon's most striking impact feature, Mare Orientale, lies on the extreme eastern limb and is only half visible when that limb is tilted most toward earth. The Galileo photos of it from 1991 were quite spectacular. All total, we get to see about 59 % of the moon's surface from earth, if we are patient enough to watch the moon for several years.

Why does the moon keep one face toward us? As Zeilik shows us on page 185, the moon is lop-sided, with the heavy hemisphere tidally fixed toward us. This is the rule; only Saturn's odd-ball moonlet, Hyperion, does not revolve around its home planet with one side rotating to stay facing its planet; it has been suggested this hamburger shaped satellite is a comet, so recently captured by Saturn that it has not settled down into a synchronous rotation.

Why then does the moon show us this tilting, or libration? As Zeilik notes, the moon's orbit is fairly elliptical, and the moon will speed up as it approaches perigee, and slow down as it climbs out to apogee. The net effect is like you shaking your head "no," tilting left and right as the speed changes. As you might remember from the discussion of eclipses, the moon's orbit also takes it 5 degrees above or below the ecliptic, so sometimes we can peer over the moon's north pole, then at other librations, look under its south pole. The effect here is similar to our head nodding "yes," up and down.

Did the Moon Rock the Cradle?

We now know that tides play a far more important role in planetary evolution than anyone could have imagined a generation ago. It is tidal stresses that heat Io's and Triton's volcanoes, melt Europa's oceans, and wrinkle the grooves on the crust of Ganymede, Enceladus, and Ariel. In our past, the moon was orbiting far closer to us, and both its tidal effects on us and ours on the moon were greater than at present. Perhaps these tides may have been vital to the origin of life of earth.

Tidal pools are one of the prime sites that biologists consider for origin of life. In the constant mixing, wetting, and drying, one can easily imagine the synthesis of complex organic molecules and the colloids which would evolve into the cell membrane to hold these early nuclei. But without a large and nearby satellite like our moon (a rarity in the solar system--only the Pluto-Charon system approximates our own), would life have arisen as fast? If the Mars fossils are real, this no longer seems as likely, for Mar's tiny moonlets would have had little tidal impact on the deposition of the carbonates and possible fossils in the Allen Hills Martian meteorite.

Mascons, Impacts, and Plumes

The presence of dense materials beneath the mare in the form of mascons could be reasonably explained by the presence of dense asteroidal bodies, still imbedded in the moon's upper mantle. It was their impacts which broke the lunar crust, and their impact energy which helped melt the upper mantle and thus led to the flooding of the basins they made. But the moon cooled fast, and at the time these impacts, was already too rigid to allow the dense bodies to sink to the moon's core. Thus as the Apollo craft orbited above them, their higher density pulled the craft lower.

Certainly such impacts happened elsewhere; all the other terrestrial are bigger targets. But if these larger bodies were still differentiating, and molten inside, the impacting asteroids would gradually sink toward the planet's core. I think the plumes where hot magma wells upward through the mantles of earth, Venus, and Mars may be the fossil tracks laid down where such asteroids descended downward to become part of the dense, molten cores. In the November 1993 issue of Earth, an article on the Permian extinction raises the possibility that a large impact in southern Africa not only almost destroyed all life on the planet, but started the plumes that ripped apart Gondwanaland via continental drift.

DEMONSTRATIONS AND OBSERVATIONS

An Earth-Moon Model

If you use a Styrofoam ball 8" across for the earth, then the moon needs to be a ball 2" across, and about 20 feet from the earth ball; the moon orbits us at about 30 earth diameters distance, as Ptolemy found about 200 A.D. This would be a nice display for one side of your classroom, particularly if some artistic student would paint both to approximate the worlds. It will also bring home to your students just how perfect the alignments of all three bodies must be to produce lunar and solar eclipses, and thus why these are rare events.

Observing the Moon

The moon is still the most interesting body to observe with a telescope, and bright enough that at many phases it can be observed by your classes even in broad daylight. To start with, a pair of binoculars will reveal much detail even at 7x, and zoom binoculars at 30x can reveal craters less than 10 miles across, not bad from a quarter of a million miles away!

For morning classes, the best phases to observe are from third quarter to waning crescent. When the moon is about 90 degrees away from the sun (first and third quarters), contrast is helped immensely by placing a Polaroid filter over the eyepiece (a piece of old Polaroid from plastic sunglasses works quite well). Rotate the filter to darken the scattered light from the atmospheric dust as much as possible; the gain in contrast is amazing, for the sky is polarized, but the light reflecting off the moonrocks is not. In the afternoon, the waxing crescent and first quarter phases can be used to like advantage.

For night observing, try to get several telescopes at a variety of magnifications for your classes to use. For the waxing crescent just after new moon, use about 20x to show off earthshine well. Use about 60x to capture the entire disk, and perhaps 120x maximum to show craters and Mare Crisium along the terminator (the sunrise or sunset line). Note that fine detail stands out near the terminator; except for the lunar mare and rays around the youngest craters, most lunar detail is NOT well seen at full moon.

One of my favorite tricks at first quarter is to let the students go "moonwalking." Because the terminator then is perpendicular to our horizon, with an equatorially mounted motor driven scope such as a Celestron C-8, put on a fairly high power (200x works well, particularly if you have one of the newer wide field eyepieces). While the motor drive tracks the moon, show the students the location of the declination slow motion, and have them sit in a chair and observe the moon comfortably with a star diagonal tilted to match their head. As they turn the declination control, the scope will sweep from pole to pole along the moon's terminator; the effect is like orbiting just above the moon. It will be hard to get some students away from the eyepiece, so set up several scopes like this if you can. For many students, this will be their most memorable experience in your class.

If you do not already have a CCD for your scope, by all means get one (they are available for less than $300 now, at .2 lux in full color). Show your students the video image of the moon on a large monitor, and have them note and identify significant details--a great class project!

Because the full moon is so bright, you might like to use a gray filter to less its glare, even with smaller scopes. Use low enough power to see the entire disk in space--with wide angle eyepieces, the moon really seems to hang there in 3-D. Fairly high magnification binoculars or zoom binoculars can make this effect even more striking, especially if they are sturdily mounted.

Have your students note the relative ages of craters, ideally with the video image. The oldest ones in low lying areas of earthside have been flooded by lava, and have gray basaltic floors. Younger ones sit atop the older ones, or on the mare; the youngest like Tycho and Copernicus are still surrounded by bright lunar rays. If you get the moonrock samples from Houston, you will find these rays are just tiny glass beads, acting like sequins at full moon to bounce the sunlight right back at earth. In fact, use to this reflection effect, the full moon is nine times brighter than the half moon at first or third quarter phase! As the moon is constantly showered by dark meteoric dust, the rays will in time be covered in this lunar soil, or regolith; hence only the youngest craters still have visible rays, although rays were formed every time a comet or asteroid struck the lunar surface.

The Moon and Mercury Density Globes

With our 12 inch earth globe as the standard hemisphere, for the moon we will need a three inch diameter hemisphere, with much less iron than even the Mars globe. It has a bulk density of 3.3 g/cc, so this hemisphere should weigh only 364 grams, or about .8 pounds. It will be obvious to your students holding this model that the moon is made mainly of silicates, and has a small iron core, perhaps accounting for its lack of magnetism.

By contrast, the Mercury globe will be only slightly larger, yet quite a bit heavier. At 5.5 g/cc, Mercury is about 67 % denser than the Moon. Our Mercury globe is about four inches in diameter, or a third the size of the Earth globe. It will take a good bit of iron to make this hemisphere weigh 1.4 kilograms, or about 3.2 pounds. The large contrast between it and the moon globe will tell your students that Mercury's core is rich in iron, and also help explain the presence of the only other magnetic field yet detected among the terrestrial.

ANSWERS TO STUDY EXERCISES

1. The moon's surface is barren, with no plant or animal life. There is no water on the surface, and no atmosphere, so one does not see the effects of water and wind erosion on the surface--only a fall of meteor dust changes the surface slowly over time. The sky is dark, even in daytime, with stars visible; they shine steadily, without the atmospheric twinkling created by our air. Craters cover the landscape, and low mountains may be seen in the distance.

2. Mercury's surface is that of a scorched desert, with no water of forms of life. Many craters and some smooth plains might be seen. Tall scarps stretching to the horizon may be in view, but no mountains as you might see on earth. The sky would be dark, with the sun, more than twice its size as seen on earth, blazing in a dark daytime sky.

3. The moon's average density is similar to the density of the rocks brought back by the Apollo astronauts from its surface. This near equality implies that the moon's composition is probably very uniform. A dense metallic core like ours would require that the mean density be much greater than

the density of the surface material. Also, a nickel-iron core could give rise to a magnetic field greater than found on the moon. Mercury's density is similar to ours, yet its surface appears similar to the moon's. Thus like us Mercury must have a dense core.

4. In both cases, comet and asteroid impacts created almost all the craters. First, the craters appear similar to the Meteor Crater in Arizona and others on earth known to be formed by impacts. Second, the amount of material in the walls of the crater or lying around it is generally almost sufficient to fill back in the crater itself, suggesting the craters were excavated by the explosion of impacting bodies. Third, the patterns of bright rays and material around the craters resembles that expected from an impact rather than volcanic ejecta. Fourth, shock waves are often visible just outside the crater (particularly in the case of Copernicus and the Caloris basin). Lastly, ejecta often forms chains of secondary craters, again quite notable at Copernicus.

5. Radioactive dating of rocks from Apollo missions shows the highlands rocks the oldest (between 4.48 and 4.0 billion years) while maria material dates from 3.9 to 3.1 billion years. The scarcity of craters in the maria also shows they formed after most of the debris had already been swept up in our vicinity; the lava flow filling in the earlier record of cratering.

6. Answers will vary greatly, but some good points to consider are:
Fission theory -- the moon is about the same density as our crustal rocks.
Capture theory -- the moon orbits within five degrees of the ecliptic.
Condensation theory -- the similar density of moon and earth crustal rocks, and with some significant differences, similar compositions. Also, this pattern seems followed by most other planets, such as the satellites of Jupiter.
Big Impact theory -- the violence of the impact drove off the light volatiles, leaving the moon very dry. The densest materials fell back to earth, leaving the moon low in density. The impact happened in the ecliptic plane, accounting for the moon's present orbit. The impact on earth was oblique, tilting us 23.5 degrees off the ecliptic in our rotation.

7. Orientale and Caloris are similar sized basins consisting of dark lava-covered central floors, surrounded by concentric rings of mountains, and surrounding regions blanketed with ejecta. They are probably the result of the impacts of asteroids of similar mass (perhaps similar to Phobos orbiting Mars) and size (about fifteen kilometers). The floor of Caloris appears wrinkled, perhaps due to rapid cooling of the melted rock.

8. Mercury's surface has scarps from the core contracting, and fewer small craters and more extensive lava flows than seen on the moon.

9. Both have weaker surface gravities than the earth. Also, Mercury is exposed to far more solar radiation, which tends to strip away gas molecules into space faster than here at earth. The helium at Mercury comes from the solar wind, and the sodium outgasses from exposed rocks.

10. Mercury is expected to have a cool, metallic core, while the moon may have a warmer rocky core.

11. The evolution of the surfaces of both bodies followed a similar sequence, based on their similar present appearances.

12. If an area devoid of craters were found, it must be far younger than the rest of Mercury already mapped. This is not unprecedented; one hemisphere of Saturn's moon Enceladus is very old and heavily cratered, but the other much smoother and younger.

13. The side facing the moon is subject to the most tidal stress, the one opposite it the least; the water is pulled from the area in between, where observers would see the moon just rising.

14. The earth pulls more strongly on the one closer to it, pulling them apart.

ANSWERS TO PROBLEMS & ACTIVITIES

1. Again, $g = GM/R^2$, so for the moon's radius and mass, we get the moon's surface gravity to be g $= 6.67 \times 10^{-11} \times (7.2 \times 10^{22}$ kg$) / (1.738 \times 10^6$ m$)^2 = 1.6$ m/s^2, or one-sixth ours.
For Mercury, $g = 6.67 \times 10^{-11} \times (3.3 \times 10^{23}$ kg$) / (2.440 \times 10^6$ m$)^2 = 3.7$ m/s^2, or about twice the moon's and a third the earth's surface gravity.

2. For the moon, $v_{esc} = (2GM/R)^{.5} = (2 \times 6.67 \times 10^{-11} \times 7.2 \times 10^{22}$ kg $/ 1.738 \times 10^6)^{.5} =$
2.35 km/sec, about one fifth the earth's escape velocity of 11 km/sec. For Mercury, the escape velocity $= (2 \times 3.3 \times 10^{23}$ kg $\times 6.67 \times 10^{-11} / 2.440 \times 10^6) = 4.2$ km/sec, or 1.75 times stronger than the moon's escape velocity.

3. Using the data from #2 on the escape velocities, Mercury's escape velocity is 4.2/2.4 = 1.75x faster than the moons. As the kinetic energy is given by $.5mv^2$, then as nothing is changed except the velocities, we find the body striking Mercury has $(1.75)^2 = 3.1$ times the kinetic energy of the same mass hitting the moon. Mercury should have larger craters than the moon, given the same size bodies impacted both surfaces.

4. Eight arc seconds is 8/3600 = .002222 degrees, thus sin .002222 = .0000388, or inverting this, we find a size to distance ratio of 1/0.0000388 = 25,783X. If Mercury 4,880 km in diameter, then it must have been 25,783 x 4,880 = 125,820,000 km or .83 AU.

5. As the tidal force is proportional to $1/R^3$, at twice its present distance the moon would only have $(.5)^2 =$ one eighth its present effect. This goes a long way to explain why the much closer but far less massive moon plays a larger role in our tides than does the distant sun.

CHAPTER 10

Venus and Mars: Evolved Worlds

CHAPTER OUTLINE

Central Question: What forces have shaped the evolution of Venus and mars, planets with hot interiors and substantial atmospheres?

10.1   Venus: Orbital and Physical Characteristics
         A. Revolution and rotation
         B. Size, mass, and density
         C. Magnetic field

10.2   Interlude: The Doppler shift
         A. Concept
         B. Application to planets

10.3   The Atmosphere of Venus
         A. Clouds
         B. Atmosphere and surface temperature

10.4   The Active Surface of Venus
         A. Highlands and lowlands
         B. Volcanoes and tectonic activity
         C. Magellan at Venus
         D. The evolution of Venus

10.5   Mars: General Characteristics
         A. Orbit and day
         B. Size, mass, and density
         C. Magnetic field
         D. Atmosphere and Surface Temperature

10.6   Remote Sensing of the Martian Surface
         A. The sands of Mars
         B. Global dust storms
         C. Canals and the polar caps

10.7   The Martian Surface Close-up
        A. General surface features
        B. Arroyos and outflow channels
        C. Volcanoes
        D. Cratered southern hemisphere
        E. The evolution of Mars

10.8   The Moons of Mars

Enrichment Focus 10.1: The Doppler Shift

CHAPTER OVERVIEW

Teacher's Notes

For a summary of the physical characteristics of the planets, refer to Appendix B in the text. A large portion of this chapter is descriptive in nature, and the current descriptions have been provided by space probes to the planets--back these up with photographs taken by the space probes and orbiters. A wide selection of such photographs is available, as are slides from them.

Distance and sizes of the planets themselves and of features are always difficult to visualize in numerical form. It's a good idea to place these in perspective; refer to the earth and features on the earth for comparison. That's really the key point of this part: that the earth forms the basis for the comparative evolution of the planets.

The discussion of Mars in the text will probably prompt questions about the Viking life-search experiments and possibilities of life on other planets. Some mention is made in the text of the life zone or habitable zone around the sun, and the ability of the planet's environments to support life. At this point, keep discussion of these ideas or speculations concentrated on the physical characteristics of the terrestrial planets. The subject of extraterrestrial life, or exobiology, is discussed in more detail in Chapter 22. If you just can't hold your students back, have them read section 22.4 now. You can return to and review these ideas later in the context of life in the Galaxy.

10.1 Venus: Orbital and Physical Properties

These are the two planets most frequently compared to earth, yet each, driven by similar processes, has ended up a very different world in many respects. Their atmospheres are quite different from ours. Both show more surviving evidence of the age of cratering. Both may have had seas of liquid water early on, but for very different reasons, both lost them. Comparative planetology compares the evolution of other worlds with our earth; for these worlds, the similarities and differences are particularly interesting and important.

Due to Venus' very slow rotation, Venus is the one place we know of where the day (243 earth day rotation) takes longer than the year (only 225 days to orbit the sun.) The combination of the two motions means the sun is above the Venusian horizon for half of the 117 day period or about 58.5 days of continuous sunlight. The cloud cover blocks most light, however, so actually the surface is about as bright as a day of heavy overcast on earth, or about 1/100th as bright as on a clear day.

No other two planets in the solar system are as similar as earth and Venus. If Venus were where the earth is now, very likely there would be life on it also. Yet for a planet as iron rich as our own, Venus' lack of a magnetic field is most perplexing. The dynamo model can not be successfully applied to any other planet! Some have suggested we have caught the lady with her field down; Venus may be in the middle of a field reversal, and the field may soon grow back. If so, continuing magnetometer observations should reveal these changes. Also, sample returns from Venus' surface might show magnetic alignments in iron-rich basalts from most of the surface. No such returns are planned as yet, although the technology to do so exists now.

10.2 The Doppler Shift

In helping my students remember the doppler shift, remind them that a red shift means the object is receding--keep the r's together. Also, if we observe a longer wavelength than expected, it means the distance to the object is also growing larger as well. In helping students remember radial velocity, I compare it to the motion of a ray of light, in our line of sight. When expressing radial velocities, in tables of star motions a negative value means the speed it is approaching us, while a positive value is a recession velocity.

Ask your local airport if they have a doppler radar yet; the FAA is now installing them in practically all major airports. Ask students what would be the significant safety advantage to air traffic controllers (and everyone up in the air) of doppler radars; perhaps some will grasp that these radars permit not only the location of the plane, but also by knowing its velocity, prediction of its future course. With sophisticated computers, collisions can be predicted far enough in advance that adequate warnings could be given and almost all of them could be avoided.

10.3 The Atmosphere of Venus

The Greenhouse Effect does make Venus a hell, but remember that it helps make the earth a heaven, too. We have just the right amount of greenhouse gases to keep the earth's oceans from freezing over; without our own carbon dioxide and water vapor, a transparent atmosphere of pure nitrogen and oxygen would allow too much solar energy to be reflected back out into space, and our world would be as frozen as Mars is now. It is this delicate balance which many ecologists fear will be threatened by human burning of fossil fuels and destruction of rain forests and pollution of the oceans. Some of the activities increase the carbon dioxide content, while others would reduce it; if either gets out of hand, life on earth would become much more difficult.

The geologic record, however, reveals sudden and large naturally occurring changes in carbon dioxide concentration in the fairly recent past. These are quite evident in new research on the ice cores from Greenland and Antarctica. Human technology had nothing to do with these changes, so perhaps volcanic eruptions pumped much new carbon dioxide out at that time.

Here on the earth, our atmospheric circulation trails behind the rapid rotation of the solid crust beneath it; this in turn sets up the coriolis effect and the motion of weather fronts from west to east across the United States. Note that this is not true on Venus; the high altitude cloud deck is circling Venus in only four days, or moving sixty times faster than the very slowly rotating crust beneath it. Apparently there is some atmospheric discontinuity, perhaps at about 30 kilometers up, that separates the rapid circulation and high winds above, with showers of sulfuric acid droplets, from the calm, clear, stiflingly hot lower regions of the Venusian atmosphere.

In considering Venusian weather, sulfuric acid plays a similar role there to water in our atmosphere. It condenses from vapor into clouds, and accretes into raindrops. Note that it is far too hot at the surface for even sulfuric acid to exist as a liquid. If it could, Venus would have lakes or even seas of liquid sulfuric acid. This acid rain falls from the clouds, but probably flashes back up as vapor when it reaches about 25 kilometers altitude. From there on down, the Russian Venera landers gave a monotonous forecast: overcast but with no wind or precipitation on the hellish surface. Yet the newest radar images from Magellan do show areas of the Venusian surface apparently swept by wind erosion; sudden changes in pressure, perhaps due to large asteroid impact or volcanic eruptions may temporarily stir up the lower atmosphere.

A major surprise in 1996 was the discovery that several rare and exotic heavy metals had "snowed" out on the top of Venus' highest peaks, where the temperatures were a little lower than the hotter surface. The metal deposits were picked up from the Magellan data due to the metals being very good reflectors of radio and radar waves.

10.4   The Active Surface of Venus

In looking at the Russian Venera lander photos, note the gray color of the basaltic flows. Basalt is the chief surface rock of most terrestrial bodies, covering our ocean basins, the dark lava flows of the lunar mare and Martian northern hemisphere as well.

In comparing the geologies of earth, Venus, and Mars, note to the students that Venus is the flattest of the three. It does have a few high volcanoes, like Maxwell Mons, yet the continents of Venus are only about two miles higher than the basins below, compared to about four-five miles between earth's continents and ocean basins. Venus has even more basin area than earth. Here we have about 70 % ocean basins, and 30 % uplifted continents; on Venus, it is about 90 % basin, and 10 % elevated regions.

Stress to the students that the thick Venusian atmosphere completely burns up meteors far larger than most which reach earth as meteorites. There is no impact crater on Venus as small as our famed Meteor Crater in Arizona, portrayed so vividly at the end of "Starman." Only the really big asteroids can pierce this dense atmosphere intact, and no craters less than about six miles across have been found by Magellan, with its initial survey now complete.

Even if Magellan has not over several months spotted changes in lava flows, proving continuing volcanism exists, strong evidence for this came in 1978 when the Pioneer Venus Orbiter began returning data. It showed the concentration of sulfur dioxide in the Venusian atmosphere quite high, but found it to be declining ever since. Since this is a very common product of volcanic eruptions, it seems a large eruption occurred in 1977, but with no others on the same scale since.

Since Venus has a similar bulk density, it is reasonable that it has a similar amount of radioactive U 238 in its core, so tectonic activity on a similar scale to our own is reasonable. But on Venus, the number of plumes or hot spots seems greater, and the mobile plates consequently smaller. Compared to earth, you might get more quakes on Venus, but of a gentler intensity than on earth. But keeping a working seismograph on the surface of this hellish planet will require much more heat resistant electronics than we presently have.

The Magellan data suggests that Venus has periods of great volcanic activity, followed by periods of hundreds of millions of years with little crustal renewal. The uniform distribution and ages of the meteor craters in the data is the basis of this tectonic behavior, very different from ours.

10.5   Mars: General Characteristics

Again planetary magnetism puzzles us. We explained the lack of magnetic field at Venus as due to its slow rotation. Yet Mars has a day only 40 minutes longer than our own, and as at Venus, no detectable magnetic field either. Challenge your students--the first to come up with workable model for planetary magnetism which can be applied to all terrestrial planets wins a free trip to Stockholm, with a few million in cash and a nice little Nobel Prize in Physics.

Note that both Mars and Venus have atmospheres dominated by carbon dioxide. Ours would be similar, except for the removal of almost all our carbon dioxide by photosynthesis and carbonate rock formation, as discussed in chapter four. On Venus, continuing volcanic activity keeps pumping more carbon dioxide into its atmosphere; but with Mars' volcanism mostly a thing of the past, this replenishment has basically stopped. Mars, with a lower surface gravity, is losing its atmosphere into space, particularly its precious water.

Compare the wide variation in surface temperatures on Mars (over 100 degrees K, or almost 200 degrees F between day and night) with our earth's much more moderate ranges (typically about 20-30 degrees F near the coasts, a little wider inland). Our oceans and atmosphere certain make the earth a far more hospitable place than present-day Mars. This wide range of temperatures will become even more notable when we discuss Mercury and the Moon, with even thinner atmospheres, in the next chapter.

10.6   Remote Sensing and the Martian Surface

It is notable that as on earth, much water of hydration is tied up in crustal rocks; the note on the red color of the sands implies that water, as well as oxygen, was present to cause the iron minerals to turn red. If we could heat these rocks, usable quantities of liquid water for colonization might be made available. The hydrologic survey instruments aboard Mars Observer were to give much more information about how much water Mars still possesses, and in what form we can find it. I personally hope a second Observer can be constructed out of spare parts and sent to the Red Planet as soon as possible to continue this research.

Mars has two very different polar caps. The winter cap may extend to about 40 degrees latitude in both hemispheres and is a thin frosting of dry ice, as shown by some Viking II lander photos. This cap sublimes (this is dry ice, frozen carbon dioxide, so it passes directly from solid to gaseous phases) rapidly in the spring, and the cap retreats noticeably week after week. But even in mid-summer, the water ice cap remains frozen; on Mars, water, like carbon dioxide, can not exist at surface pressures presently as a liquid.

10.7   The Martian Surface Close-up

It is obvious from the eroded channels and outflows that once Mars had considerable quantities of an abundant running fluid, presumably water. Some geologists speculate that hydrogen sulfide might have instead been the fluid, but most believe water was abundant as a liquid in a warmer, wetter Mars of a billion years ago. If Mars' volcanoes were much more active then, their outgassing would have thickened the atmosphere (even twice its present density would have been enough) and added such steam to the clouds, leading to rainfall and erosion. Even if the heating had not been active volcanism, underground it could have melted the tundra, with geysers erupting at the surface and creating localized mudflows and channels.

Some JLP geologists think Mars was even wetter than these localized patterns indicate. According to them, the huge expanses of uncratered terrain over much of the northern hemisphere are in fact due to the areas being covered by sedimentary rocks, formed in a vast ocean that covered much of the northern hemisphere and low lying areas all over the red planet. In support of this, they point to numerous examples of terraced deposits, apparently formed by ancient sea level stands and wave-worn beaches. The carbonates in the Allen Hills meteorite are also formed under water. Exactly how extensive these oceans were may be revealed by a replacement Mars Observer, or by proposed Russian missions later in the decade.

10.8 The Moons of Mars

Deimos and Phobos are probably captured asteroids and as such, give us an interesting glimpse of the smaller bodies found just beyond Mars' orbit. It is notable that both are much darker than a terrestrial planet, including the dark basaltic flows on the moon. This of course is due to an abundance of black carbon on their surfaces. Surfaces with such low albedos are in fact common in the outer solar system.

The reason carbon black is generally lacking in the inner solar system is the lack of hydrogen close to the sun. If there is a lot of hydrogen around, almost all the extra oxygen is quickly turned into water, very abundant in the outer solar system as ice. But if a surplus of oxygen persists, it oxidizes the exposed carbon into the familiar forms we see around us in the inner solar system, carbon dioxide and carbonate rocks, like limestone.

Note that both Martian moonlets are fine examples of non-differentiated bodies. Neither is anywhere close to being spherical in geometry, and both have heavily cratered surfaces. It is true that Phobos has some interesting-looking parallel "stretch marks" on its surface; the origin of these is still a mystery, but it may be related to the tidal forces that will play such a major role in explaining the surfaces of our moon and the moons of the jovian planets.

There is reason to think the orbit of Phobos is decaying and the moonlet is approaching Mars Roche Limit, where the tidal stresses will become so great as to break apart the already stressed body. If this occurs, then the pieces will grind together, ultimately making a ring of fine dust around the Martian equator. This is likely the process that produced the rings around the jovian planets. If Mars' tidal stress is not great enough to break Phobos apart, it may continue its downward spiral and hit Mars with a grazing impact, making an elongated crater on the Martian equator. Several such craters have been noted there, so this may have happened before.

ADDITIONAL RESOURCES

Transparencies 10.2, 10.4abc, 10.5, 10.16, 10.28, and F.8, as well as slides of Valles Marineris (chapter cover) and 10.18 will assist your presentation of this chapter.

The two videos mentioned for Chapter 4 both fit well into this chapter. In addition, Sagan's Cosmos 105, "Blues for a Red Planet", is a good supplement to the Mars discussion. Nova's "Venus Unveiled" in 1996 is an excellent report on the Magellan results..

If you can afford it, Sky Publishing has produced a very accurate Mars and Venus globes for about a hundred dollars. The look great in the classroom, and the Mars globe can be a real asset if you want to observe Mars telescopically and show your students which hemisphere of Mars they are observing. Both Sky and Telescope and Astronomy have tables to help you near oppositions.

SUPPLEMENTARY ARTICLES

1. "Digging Deeper into Life on Mars," Astronomy, April 1988, page 6.
2. "Modeling Terrains around the Solar System," Astronomy, March 1989, page 73.
3. "Searching for the Waters of Mars," Astronomy, August 1989, page 20.
4. "New Russian Views of Mars and Phobos," Astronomy, September 1989, page 28.
5. "Surveying Ancient Martian Floods," Astronomy, October 1989, page 50.
6. "Magellan Unveils Venus," Astronomy, February 1991, page 44.
7. "Rendevouz with Venus," Astronomy, April 1991, page 38.
8. "Venus, Planet of Fire," Astronomy, September 1991, page 32.
9. "Mars Observer: Return to the Red Planet," Astronomy, September 1992, page 28.
10. "Observing the New Mars," Astronomy, November 1992, page 74.
11. "The Ice Ages of Mars," Astronomy, December 1992, page 40.
12. "What makes Venus Erupt," Astronomy, January 1993, page 40.
13. "The Sands of Mars," Astronomy, June 1993, page 26.
14. "Mars: Did It Once Have Life?," Astronomy, September 1993, page 26.
15. "Mars: The Russian Plans," Astronomy, October 1993, page 26.
16. "the Odd Little Moons of Mars," Astronomy, December 1993, page 48.
17. "To Boldly Go (robot explorers)," Astronomy, December 1994, page 34.
18. "Magellan Reveals Venus," Astronomy, February 1995, page 32.
19. "Messengers from Mars: Meteorites from the Red Planet," Astronomy, July 1995, page 44.
20. "New Discoveries: NASA's Discovery Program," Astronomy, November 1995, page 36.
21. "The Mars that Never Was," Astronomy, December 1995, page 36.

ADDITIONAL DISCUSSION TOPICS

A Venus Balloon

Since no man-made lander has survived for more than about an hour on the surface of hellish Venus, perhaps a good follow-up to Magellan in the late 1990's might be a Venus balloon, with instruments and cameras in its gondola, floating beneath the rain of sulfuric acid droplets, yet high enough up to avoid the excessive heating of the lower regions of the Venusian greenhouse; perhaps an altitude of 20-25 kilometers might be optimum. Once the observers on earth were satisfied with their survey of one area, the balloon would ascend just enough to pick up some breezes to blow it to a different area for observation.

A Mars Balloon and Rover

Plans for new exploration of Mars are much farther along than the Venus concept above. The Planetary Society, along with the French and Russian space programs and NASA, has in 1992 actually tested a balloon borne instrument probe in the Marslike deserts of the American Southwest and on Russia's rugged Siberian volcanic fields. It would descend to the surface of Mars as the temperature dropped every evening, to gather surface data, then be carried aloft as the sun warmed the black balloon the next morning.

The Russians and NASA also have Mars Roving Vehicles well along in testing, with our Mars Surveyor to be sent to the Red Planet in 1997. The possibility of past life on Mars will probably lead to funding of a variety of small missions to the Red Planet in the next few years.

The Martian Micro Fossils

In what some are claiming is the greatest headline of the age, on August 8, 1996, a NASA team of scientists announced they had found much evidence of microfossils in a carbonate rock originally from Mars. This is taken from a article written for the October 1996 issue of <u>Sky and Telescope</u> magazine, and available on the Internet at http://www.skypub.com/news/marslife.html.

J. Kelley Beatty writes the meteorite, identification ALH84001, was blasted off the Martian surface about 16 million years ago, then fell to Earth in the Allan Hills of Antarctica about 13,000 year ago, to be collected by NASA scientists in 1984.

For a decade, it was wrongly identified as a possible piece of the asteroid Vesta, but discovery of carbonate deposits and its great age (4.5 billion years old) indicated it was part of Mars' original crust (much older than ours, for smaller Mars cooled faster).

Microscopic examination of the carbonates showed rod shaped objects, smaller but otherwise reminiscent of the oldest fossils on Earth, about 3.45 billion years old. Also, the pattern of iron mineralization around the rods is very reminiscent of the patterns laid down by anaerobic bacteria on earth even now. Lastly, the complex PAH (polycyclic aromatic hydrocarbons) are typical of the decay products of simple life on earth.

Hello, Grandaddy?

If these microfossils are real, and significantly older than larger but similar ones on earth, **could we be their children?** As Mars cooled faster, life could have begun there perhaps half a billion years before conditions on earth were habitable. As collisions with comets and asteroids were much more common back then, the chances of a viable piece of Martian "live rock" making it to earth just in time to seed our condensing oceans perhaps 3.6 billion years ago should now be considered. Remember live bacteria were brought back from the Surveyor that had sat a decade on the Moon by the Apollo 14 mission in the 1970's, so stranger things have happened!

DEMONSTRATIONS AND OBSERVATIONS

Venus' Phases

Any small scope, such as a 60mm refractor at 50x, is adequate for showing the students the phases of Venus. Unless Venus is near conjunction with the sun, this can even be done in the daytime. Use the planetary locations published monthly in <u>Astronomy</u> or <u>Sky & Telescope</u> to find were Venus is in relation to the sun. Then knowing the field of view of your binoculars (typically about 7 degrees for 7x binocs), and AVOIDING looking at the sun directly (stand in the shadow of a building for better contrast), search that area of the sky, looking for a tiny, bright dot of light, also metallic in its high albedo. Many aviators in war have tried to shoot down Venus, mistaking it for a hostile aircraft. If the day is clear enough, once you have located it, most of your students can probably see it with the naked eye in broad daylight. Try walking to where Venus appears to sit atop a convenient tree or at the edge of a building to help your students locate it.

Once you have found Venus, then view it at about 50x with your scope. If it is just coming from superior conjunction, behind the sun, it will appear tiny (it is on the far side of its orbit, almost 2 A.U.s distant) and almost fully lit. As it overtakes the earth in the coming months, watch the disk grow larger, but less fully lit.

When farthest from the sun at greatest elongation, it is 47 degrees from the sun; at greatest eastern elongation, it then sets three hours after the sun in the evening sky. At this point in its orbit, it appears half lit.

During the next two months after greatest eastern elongation, Venus rapidly overtakes the earth and appears to retrograde. Telescopically, the disk grows larger and the crescent becomes dramatically thinner day by day. A few weeks before and after inferior conjunction, when Venus passes between us and the sun, the crescent Venus can be noted even with 7x binoculars; so people with extremely good eyes even claim to note it visually. Galileo first observed these changes in 1610-1611 and used them as powerful proof that Venus was independently orbiting the sun, not on Ptolemy's epicycle. In Ptolemy's model Venus would have always lain between us and the sun, so always appeared as a crescent. The fact that Venus shows all phases also shows it revolves all the way around the sun.

Past showing your students Venus' phases, it is doubtful that you can spot any additional detail in the clouds of our mysterious sister planet. Because Venus is so bright, it is best to observe it in the daytime, or at least at twilight, when it is highest in the western sky and the bright sky background doesn't make Venus appear so glaring. As Venus gets lower in the sky, note the atmosphere works like a prism; the top of Venus appears bluish, and its bottom red. This is not chromatic aberration in your telescope, by the way.

You might try using a violet filter on the eyepiece; sometimes elusive cloud formations stand out better in the shorter wavelengths. The spectacular photos from Pioneer Venus Orbiter were all made in the ultraviolet, for instance; you can not hope for anything like that detail from the earth.

One interesting phenomena you might glimpse, however, is the elusive ashen light. This appears much like earthshine, with the night side of Venus glowing faintly behind a brightly lit slender crescent. But earthshine is due to light bouncing off earth back up to the moon, while the ashen light probably is due to a planet-wide auroral display. With no Venusian magnetic field, charged particles are not focused toward the poles, as with our auroras.

Observing the Red Planet

While Venus showed only phases, telescopic observers of Mars using larger scopes can see much more detail. At oppositions, such as in the spring of 1993 and early summer of 1995, the earth draws close enough to Mars that it becomes the fifth brightest object in the sky (after the sun, moon, Venus, and Jupiter) and the orange red color stands out well. High power (200x or greater, if atmospheric stability allows) and an orange or red filter are recommended for bringing out elusive surface details on Mars.

Look to the poles for the polar caps. If it is Martian spring, the cap tilted toward earth and sun will be shrinking week by week. At the opposite hemisphere, you may see the fall caps growing, although usually this growth is veiled by a haze of clouds, snowing down the dry ice on the winter hemisphere. If it is Martian summer, the tiny residual cap of water ice which never melts is much smaller and more difficult; congratulations if you can spot it.

The overall color of Mars is orange, due to the vast expanses of rusty deserts revealed by the Viking landers. But large areas are covered by dark, basaltic lavas; these dark markings seem to change with the seasons. They were mistaken for growing vegetation by supporters of Percival Lowell's canals early in this century.

We now know these seasonal changes are due to high winds and huge dust storms which in the spring may sweep the dark flows clean and dark, then dust them in the fall to make them appear brighter. Again, try to identify features on Mars with a Mars map, and relate them to the exploration of the Red Planet in greater detail by the Russian missions over the next several years.

Observing Canals

Of course we know the canals are not there, but it is easy to demonstrate why Schiaparelli and Lowell thought they saw them. Look closely at a newspaper photo. It is really a series of dots, on microscopic scale unrelated to each other. It is only when we back off to see the whole page that the pattern of the photo is again evident to us. The human brain has a tremendous capacity to relate the unknown into a meaningful, if incorrect, pattern. All of us have in moments of distraction, stared at the pattern in wood grain, and imagined faces, shapes, even monsters coming out of the grain. The early visual observers, peering through a turbulent ocean of air, saw transient linear patterns in the dark lava flows, and the eye, brain, and hand all mislead the observers to draw these real features as much more organized, linear, and distinct than the orbiter photos would show. When you go to the eyepiece with a strong preconception of what you will see, you can often trick yourself into thinking you see what is not there.

Venus and Mars Globes

A Venus globe could be made out of the other hemisphere of the Earth globe mentioned in chapter four. As the bulk density of Venus (5.2 g/cc) is only a little less than earth's, it would weigh only about 200 grams, or a half pound less than the earth globe. Students can't probably tell the difference.

The Mars globe would need to be only half as big as your earth globe, but much less dense. Mars has a bulk density of 3.9 g/cc, or about 71 % of the earth's, so you would need a good bit less iron to accurately mix with your plaster of paris. This would also go a good way in explaining the absence of a Martian magnetic field.

Why Mars is Red

If you have some wet gray clay in your area, try bringing some to class and drying it in the sun, or even in an oven. The gray color is Prussian blue, a form of iron without much oxidation. Drying it out and exposing it to atmospheric oxygen will cause it to rust and redden day by day. The soils of Mars are highly oxidized, as Viking landers discovered in their chemical tests back in 1976.

Why No Oceans are on Mars

If your students can't grasp why water is not a liquid on Mars now, bring into the classroom a small carbon dioxide fire extinguisher. Shake it and let them hear the liquid carbon dioxide slosh around inside the cylinder. Ask them if you release the carbon dioxide, will it remain a liquid outside the container. Most will say no; you might even release just enough to form a small bit of dry ice.

Have the students watch the dry ice sublime directly into a gas. Why then, you ask, is the carbon dioxide liquid inside the cylinder? If they pick up the heavy, thick walled cylinder, they note it more than atmospheric pressure to keep carbon dioxide liquid at room temperatures. As Mars has a atmosphere only about 1 % as thick as ours, on Mars water is not going to be a liquid, either.

ANSWERS TO STUDY EXERCISES

1. The surfaces of Venus and Mars are dry and rocky. Impact craters, volcanic mountains, and rift valleys are features common to both.

2. Venus is most like the earth in diameter, density, surface gravity, and internal structure, as inferred similar sizes and densities of the two planets. Venus and earth differ most in surface temperatures and atmospheric compositions.

3. Except for the lack of large metallic core, Mars is most like the earth in composition and internal structure. Mars also has polar caps containing some water ice. Mars and earth are most different in their atmospheric composition.

4. Mars' day and axial tilt are similar to those of earth. Its surface temperatures are not as extreme as those on Venus. While conditions on Mars are not adequate to support life, they are not as extreme as on Venus. Temperatures are colder than extremes on earth in places people can live, but not intolerable with proper protection. While the atmosphere is not capable of supporting humans, it is not as corrosive as the sulfuric acid clouds of Venus. While it can have water vapor in its atmosphere and water ice at its poles, it can't have liquid water like the oceans that cover most of earth. Its polar caps are chiefly dry ice, and its atmosphere is chiefly carbon dioxide; almost all of our carbon dioxide has been cemented away downstairs in carbonate rocks.

5. Arroyos are formed by a sudden wash of water over a surface that does not absorb water well. The flow erodes the surface and forms branching channels. But water can't flow on Mars now. So, for the Martian arroyos to have formed, the planet must have had a denser atmosphere (and so higher surface pressure) with more water vapor in it, compared to the present atmosphere.

6. As on earth, volcanic activity on Mars probably added large amounts of carbon dioxide and water vapor to the Martian atmosphere in the past. Other processes have dissipated most of the gases.

7. A major constituent in the atmospheres of both Mars and Venus is carbon dioxide. The atmospheres are different, however, in that the Martian atmosphere is a very thin, low-pressure gas, while Venus' atmosphere is much denser with a high surface pressure.

8. In winter, Mars' polar caps consist chiefly of dry ice and some water ice. In summer, the dry ice has sublimed back into the atmosphere, leaving only the tiny residual water ice cap.

9. The more impact craters we see on the surface, the longer the surface has sat unchanged by internal processes and thus the older it is. In comparing the three worlds, Mars has the most craters, and thus has the least-evolved, thus oldest surface of the three. Venus has more craters on its rolling plains than the earth, but they are all large (5 kilometers or wider)--all smaller bodies burned up in Venus' thick atmosphere. The rarity of impact craters implies the earth's surface is probably the youngest and most active, although certain areas of Venus were once quite active, according to Magellan data.

10. Compared to the earth, Mars should have few and less intense quakes. Mars has a smaller mass and diameter than earth, thus lost its internal heat into space much more quickly. The volcanoes attest to the fact that the mantle was very active in the past.

11. The earth, if you were observing it from space, because it rotates the fastest of the terrestrial and has the largest radius. For earth-based observers, Mars would show the next most observable shifts.

12. As the polar regions rotate slowest, radar reflections from them should show no shift.

13. Only the earth possesses a significant magnetic field. Mars has a rotation period similar to ours, but its lower density suggests a core with much less elemental iron than earth. Venus has almost the same density and probably similar iron core as earth, but its much slower rotation doe not generate the dynamo. But if rotation is so vital, why does Mercury, also a very slowly rotating world, generate a field far stronger than Mars' or Venus'? No good explanation at present!

14. Like the ancient lunar highlands, the surface of Deimos and Phobos are heavily cratered. However, this surface is rich in carbon and much darker than the moon's volcanic plains.

ANSWERS TO PROBLEMS & ACTIVITIES

1. First find Mercury's equatorial rotation velocity. As Mercury's circumference is 15,330 km and its rotation period is 58.6 days (5.1×10^6 sec.), this means $V = 3.01$ m/sec of dividing by the speed of light, $c = 3.0 \times 10^8$ m/s, this means the doppler shift is 1×10^{-8}; for a rest frequency of 10×10^9 Hz, this means the change in frequency much be 100 Hz.

2. Comparing the data tables in the text, Mars' escape velocity is $5.1/11.2 = 0.46$ earths. For Venus, the ratio is $10.3/11.2 = 0.92$ earth's escape velocity.

3. Using Phobos, we find its period is seven hours, thirty-nine minutes (or 27,540 sec.) and its average orbital radius is 9.40×10^6 meters, we neglect the tiny mass of Phobos to find that
$M_{mars} = 4 \times (3.14)^2 \times (9.4 \times 10^6)^3 / 6.67 \times 10^{-11} \times (2.754 \times 10^4)^2 = 6.4 \times 10^{23}$ kg.

4. The minimum distance would come when Mars is at perihelion just as earth is at aphelion. Considering orbital eccentricities, we thus get earth at 1 A.U. x (1 + .017) = 1.017 A.U., and Mars at aphelion at 1.524 (1 - .093) = 1.382 A.U.; the difference is thus .3653 A.U., or converting into kilometers, about 54 million kilometers.

5. Sin x = 6,786 km / 5.1×10^7 km = .000119, so Mars subtends .00682 degrees or 24.5 arc sec.

6. Venus has a surface gravity of .91 that of earth's; Mars' surface gravity is only 39 % of the earth's.

7. As 15 arc sec = .0041667 degree, we get sin .0041667 = .0000727, so its size to distance ratio is 13,751 X; multiplying this by Mars' diameter of 6,786 km, we find Mars was 93.3 million km distant, or .622 A.U., almost twice as distant as in the much more favorably 1986 opposition.

CHAPTER 11

The Jovian Planets: Primitive Worlds

CHAPTER OUTLINE
Central Question: How do the Jovian planets differ from the terrestrial ones, and what do these differences imply about different evolutions?

11.1   Jupiter: Lord of the Heavens
      A. Physical Characteristics
      B. Atmospheric features and composition
      C. A model of the interior
      D. Magnetic field
      E. Auroras

11.2   The Many Moons and Rings of Jupiter
      A. Io
      B. Europa
      C. Ganymede
      D. Callisto
      E. Asteroidal moons
      F. The Rings of Jupiter

11.3   Saturn: Jewel of the Solar System
      A. Atmosphere and interior
      B. Similarities to Jupiter

11.4   The Moons and Rings of Saturn
      A. Titan
      B. Other moons
      C. The ring system

11.5   Uranus: The First New World
      A. Atmospheric and physical features
      B. Moons and rings
      C. Magnetic field

11.6   Neptune: Guardian of the Deep
      A. Physical properties
      B. Moons and rings
      C. Atmospheric features
      D. Magnetic field
      E. Triton

11.7   Pluto and Charon: Guardians of the Dark
        A. Orbital and physical properties
        B. Charon: Pluto's companion planet

CHAPTER OVERVIEW
Teacher's Notes

When dealing with the Jovian planets, students most often ask questions about the nature of Saturn's rings and the planet Pluto. To explain how tidal effects can either cause an object to break up or prevent a body from coalescing, you might try a qualitative explanation of the Roche limit. A laboratory exercise on the rotation of Saturn's rings, determined by measuring Doppler shirts in spectral lines, is available from Sky Publishing Company.

Again use slides from the Voyager spacecraft to illustrate recent findings about the four gas giants. Worth mentioning in detail are the four ring systems, and the degree of meteorological activity in the upper atmospheres of each; especially contrast the blandness of Uranus with the violent activity of its twin, Neptune. Each of the satellites of the Jovian had its own individual story to tell, some of them tales of great and continuing violence. The outer solar system, while cold, is far from dead. The magnetic field of all four also should be compared, noting the extreme deviations from the conventional Dynamo model found in the tilts of the fields of Uranus and Neptune.

Point out that almost all of these discoveries are recent and due to the Pioneer and Voyager missions; simplified explanations based on earth-based observations are inadequate. These space missions boosted our resolving power of the planets tremendously, and so increased our information about them and their satellites. Compare the atmospheric activity, rings systems, satellite systems, and magnetospheres of all four together.

While Pluto is still the only world not yet explored by probes, new information is coming in about it as well. The 1978 discovery of Charon allowed us to find the mass of Pluto, while the 1992 occultation showed it to be smaller and less dense than our moon, yet with a (temporary?) atmosphere near perihelion. Use the Pluto-Charon system to show how Kepler's and Newton's laws let us find the mass of the two bodies.

Some question Pluto's planetary status. Certainly its orbit is the least planetary, with a seventeen degree tilt to the ecliptic (it is presently in Coma Berenices--your birthsign?), and so eccentric that from 1979 through 1999 it is closer to the sun than Neptune. Thus many speculate that Pluto is in fact an escaped moon of Neptune. The discoveries in 1992 and 1993 of smaller versions of Pluto, even farther out than Pluto but closer to the ecliptic, makes us question the definitions of comet, asteroid, and planet. Remind the students that the sun's gravitational pull does not abruptly end at Pluto--it may control the behavior of comet nuclei over a light year, or 100,000 AU distant.

11.1   Jupiter: Lord of the Heavens

The worlds beyond the asteroid belt are very different from the five terrestrial worlds we just studied. They formed farther from the sun, in a far colder realm, so are much lower in density. The four large Jovian giants are made chiefly of the light elements, hydrogen and helium; in some cases, their atmospheres are turbulent due to internal heating. The magnetic fields have proven even more baffling than those of the terrestrial planets, for in some cases they are tilted far off the rotational axis, and do not even pass through the core of the gas giant.

Their moons are even more diverse, most made of ices, but with varying amounts of rock indicating different sites of formation and the effects of tidal heating by their home planets. On these smaller worlds, the intensity of the major collisions has also played a major role in their present appearance.

Jupiter by itself outweighs ALL the other planets COMBINED; it makes up 70 % of the mass of the known planets. Even though Saturn is almost as large, it is not anywhere near as dense as Jupiter. Still, Jupiter's bulk density is only 1.3 g/cc, slightly higher than water, or just a quarter as dense as the Earth, Venus, and Mercury.

Note the temperature gradient in the Jovian atmosphere. As most heat comes from within, the belts are the lowest and warmest of the visible cloud features; they are made of droplets of ammonium hydro sulfide. Above them the ammonia freezes into the white ice crystals which make up the zones, and the Great Red Spot is a huge high pressure dome that is the highest and coldest visible portion of the planet.

Jupiter stinks! Ammonia is familiar to anyone who has cleaned a bathroom, and hydrogen sulfide is what you get if you go Easter Egg hunting a week late. Yet it is these obviously organic compounds which make some scientists speculate that life could have arisen in the clouds of Jupiter.

In helping students grasp the concept of liquid metallic hydrogen, point to the periodic table (yes, every astronomy classroom needs one hanging on the wall!) and note that hydrogen is in group 1A, with lithium, sodium, and potassium. Anyone who has taken chemistry is probably familiar with the silvery appearance of elemental sodium. It is noteworthy that in June 1996, for the first time in history, liquid metallic hydrogen was produced in a laboratory here on earth. The observed properties were almost exactly what had been predicted.

As Zeilik notes, the dynamo seems to work well with Jupiter's magnetic field; its rapid rotation and metallic hydrogen mantle do produce a field about ten times stronger than our own. Even its magnetic declination of 10 degrees is similar to earth's. But just wait until we get to Uranus and Neptune...

11.2   The Many Moons and Rings of Jupiter

Note the trend in densities for Jupiter's four Galilean moons is the same as for the entire planetary system. The smaller, denser bodies are on the inside, closest to the source of heat; the outer bodies are colder, larger, and lower in density, because they have held more of the volatiles lost by Io and Europa.

Note the crater counts in the photos. The very young surface of Io has no craters; slightly less active Europa shows only three; while Ganymede has plenty, but with some tectonic features; and Callisto's surface is saturated with craters. Also note the relief; Io has huge volcanoes like Prometheus, while Europa is very flat (no more than 100 meters high anywhere on its moon-sized surface).

Ganymede shows only moderate relief; the sides of some grooves are as much as a kilometer higher than their valleys. The Blue Ridge Mountains and Shenandoah Valley are similar in topography to the grooves of Ganymede.

Compare the huge impact at Valhalla with similar ringed basins on the moon and Mercury. All are the result of impacts which almost broke these substantial moons apart, testimony to the violence of the age of accretion. For smaller Miranda, this may indeed have caused such a break-up.

Why can't Jupiter have a nice, bright ring system like Saturn's? Remind the students that as Sagan puts it, "Jupiter is a star that failed." Jupiter emits twice as much heat as it gets from the sun, and the rings of Jupiter orbit close enough to it that any ice would be vaporized. Only dark dust particles, best seen from behind Jupiter, can exist this close to the giant planet. Io's volcanoes are one like source of such materials, for they erupt so violently that Io's gravity can not hold them. Newest evidence suggests this in falling debris may say in the ring system for only hours or days before raining down to burn up in Jupiter's atmosphere.

11.3   Saturn: Jewel of the Solar System

While Saturn seems a more placid version of Jupiter, the Great Storm of November 1990 proves that even seemingly placid planets can have surprises. Note in the comparison of interior models that the metallic hydrogen region in Jupiter's mantle is believed to be far more extensive than at Saturn. Saturn's magnetic field is almost precisely aligned with its rotation axis--no clue yet as to the surprises we will find at Uranus and Neptune.

11.4   The Moons and Rings of Saturn

Ask your students which body in the solar system has an atmosphere most like our own in composition and pressure. Only a few sharp students will respond that it is Titan. Of course, you aren't able to breathe it, either...no oxygen and far too cold. Yet Titan's atmosphere is far more similar to earth's than is either Mars' or Venus'.

Stress to your students how different the two hemispheres of Iapetus are. One side is ten times darker than the other. When the icy hemisphere faces earth, a 4 inch reflector will show you Iapetus; several weeks later it will take a 10 inch reflector to spot the moon, now with its far darker hemisphere facing us. Such a variation is not unheard of--consider the mare or earthside face of our moon is darker than far side; Iapetus takes it to the extreme that in the original book version of "2001" the encounter with the monolith takes place there, not in the Galilean satellite system as in the movie. Clarke proposed the aliens had painted one side black just to call our attention to it.

Just how thin are Saturn's rings? If only 3 km thick, but over 300,000 km across, then to model them, you could use paper .1 mm thick and cut a circle 100 meters across to cover an entire football field! Again, note to your students how at equinoxes the thin rings practically vanish for months, as happened in 1995-6, and will next happen around 2010.

When we first saw the spokes in 1990, we soon realized they were defying gravity. As well shown in "To Boldly Go", the spokes move in unison, like the spokes of a rotating wheel. But the rings are made of billions and billions of tiny moonlets, each in its own independent orbit around Saturn. The particles closest to Saturn should orbit the fastest, and the alignment of the spoke pattern should quickly vanish. Consider the magic of the spokes to be magnetic levitation, holding the charged particles temporarily in alignment with the planet's rotating magnetic field.

11.5   Uranus: The First New World

If you do get to see a comet telescopically while it is still fairly far from the sun, Herschel's mistake will be easily understood. The coma of a comet is initially just a blue-green blob, sometimes disklike in appearance. The fact that Uranus moved much slower than even Saturn was the initial clue to its great distance and planetary status; read Carl Sagan's article in the March 1995 <u>Astronomy</u>.

Consider how odd the year is at Uranus. With its odd tilt, an observer at one pole would spend 42 years in daylight, then 42 years in darkness. Not that it would matter much; the sun's intensity at 20 AU would be 400x less than here on earth. Still, I suspect the bland appearance of Uranus as seen by Voyager in 1986 may be transitory. Atmospheric circulation may change quite a bit at the equinoxes, to next happen about 2000; in the summer of 1993 a large storm, like Saturn's of 1990, was photographed. Many scientists, looking at the featureless Voyager images, thought that earlier drawings of Uranus with prominent belts and zones were false, like Lowell's canals. I contended that as Uranus approached equinox, and more of the globe experienced both day and night every 16 hours, the atmospheric circulation would pick up. The Hubble Space Telescope images in 1994-6 have indeed proven this to be the case.

It is notable that the hood seen over the south pole in figure 11.29 is the area in constant exposure to the sun. Perhaps the solar radiation interacts with the methane there to form the same type of reddish hydrocarbons which color the high clouds of Titan.

Note in the interior model that metallic liquid hydrogen is not expected. The magnetic fields of Uranus and Neptune, like the earth's, must be derived from an iron core; the pressure in the smaller Jovian interiors is not great enough to make hydrogen act as a metal. Note also that elemental hydrogen and helium are less abundant, and icy materials more so; some scientists suggest retitling Uranus and Neptune as "Water Worlds" or "Ice Worlds" rather than classing them with very different Jupiter and Saturn as Jovian planets.

Even though Uranus is flipped over on its side (perhaps from a great collision, as suggested in "To Boldly Go"), it is notable that its moons and rings all still orbit in its equatorial plane. This makes our own moon's orbit along the ecliptic even more exceptional.

As also suggested in "To Boldly Go," Miranda's diverse surface may be the result of the moon being struck so hard that it broke into several huge fragments, at least some of which gradually fell back together, but in jumbled, haphazard fashion. Saturn's moon Mimas has a huge impact on one side which was almost able to break it apart, and Saturn's moon Tethys may have been broken into two pieces, which immediately came back together but still preserve a 300 km wide impact crater and a great crack that encircles Tethys. At Titania and Ariel, some tectonic activity created great rifts, and while few believed the tall mountain on Oberon could be a volcano in 1986, the subsequent discoveries at Triton make this now a real possibility.

The rings of both Uranus and Neptune are very dark, being made of carbon-rich particles much like cometary debris. The contrast in albedos is great--only 5 % for the less well-known rings around Uranus and Neptune, versus 80 % for the famous and easily visible system encircling Saturn and making it the jewel of the solar system.

Now that you have won your Nobel prize for explaining the fields around all the terrestrial worlds, let's try something a little tougher. While the other magnetic fields were within about 10 degrees or so of the rotation axes, those of both Uranus (59 degrees) and Neptune (47 degrees) are far off the mark called for by the dynamo theory. Things are even more complicated by the fact that in neither case does the magnetic field even appear to pass through its supposed source, the planet's core. It has been suggested that the huge oceans of water that make up the mantles of these green giants might, with ammonia in solution, become the electrically conductive fluid materials to make the dynamo work our here. Perhaps some eddy currents in these oceans account for the strange magnetic field. If so, it might change over short periods of time--orbiters, anyone?

11.6   Neptune: Guardian of the Deep

As bland as Uranus was, no one expected much weather in the clouds of Neptune. But as Voyager II neared Neptune in July 1989, it revealed a far more interesting disk. There were deep blue belts and lighter zones, and a huge Great Dark Spot at the same latitude and only a little smaller than Jupiter's Great Red Spot. Neptune is denser than Uranus, so it must have more heating from gravitational contraction than does Uranus'. For whatever reason, the atmosphere of Neptune is much more turbulent in appearance than Saturn's, and a delight to meteorologists. The Hubble Space Telescope data has clouded the picture more, for in 1995-6 the previously striking Great Dark Spot of Neptune has vanished, to be replaced by a similar one in the opposite hemisphere!

As mentioned in chapter four, the discovery of Neptune is a great yarn itself. Look for good accounts of it in the September 1996 issues of both <u>Astronomy</u> and <u>Sky and Telescope</u>, which were timed to commemorate the 150th anniversary of Galle's recovery of the eight planet.

The clumpiness of the rings at Neptune were confusing; they were initially thought to be just "ring arcs" rather than the complete but clumpy rings revealed upon closer examination. Both Uranus' and Neptune's rings were detected via occultations of stars by observers on earth. The 1977 data on Uranus revealed the nine densest rings much as shown a decade later by Voyager II, but the sketchy data on Neptune was confused by the fact that the clumpy rings occulted the star on one side of Neptune, but thinned out so such that no dimming of star light was noted on the other side.

Triton is a fascinating world, as the photo on p. 240 reveals. It is the only large moon to orbit its planet retrograde, backwards compared to the planet's rotation. This helps explain the tidal stress that led to nitrogen geysers well shown in " To Boldly Go." At Triton, nitrogen takes on the role that water does on earth and methane does at Titan; it is at its triple point, and exists as the chief gas in the thin atmosphere of Triton, lakes and geysers of liquid nitrogen on Triton's surface, and frozen as thin ice clouds or glaciers on the surface.

11.7   Pluto and Charon: Guardians of the Dark

The orbit of Pluto is far and away the strangest of any planet, if indeed it deserves that title. It is much more like those of the short period comets, for its perihelion in 1989 placed it about 30 AU, for two decades inside the orbit of Neptune. Yet its aphelion in 2113 will find it out at 50 AU, two billion miles beyond the orbit of Neptune.

If viewed from the top, one would think the crossing orbits of Neptune and Pluto (in 1979 and 1999, for instance) might eventually lead to a collision, such can not happen. Pluto's orbit is comet-like as well in its avoidance of the ecliptic. It is tilted some 17 degrees, so in 1989 it lay well above the ecliptic, and will always lie several million miles outside Neptune's orbit when Pluto crosses the ecliptic. Thus Pluto can visit some distinctly non-zodiacal constellations, including Coma Bernices where Pluto is presently.

Even Pluto's atmosphere is probably temporary. At perihelion the solar heating would be great enough to vaporize some of the methane ice on Pluto's surface, yet freezes solid when Pluto pulls out to aphelion. In this respect also, Pluto should be considered a large comet, just bigger than Chiron and other cometary nuclei known at present. There may be thousands of other ice worlds like these still farther out, just waiting for more extensive surveys to reveal them--one candidate was found in November 1992, and as of August 1996, at least 26 of the "Kuiper Belt" bodies have been discovered, with fairly circular orbits between 33 and 38 AU from the Sun.

101

ADDITIONAL RESOURCES

Slide 11.10 and transparencies 11.3, 11.6, 11.7, 11.22, 11.27, 11.31, 11.35, 11.39, 11.41abc, and 11.47 can supplement your chapter presentation well.

Patrick Stewart (Jon Luc Picard of Star Trek, TNG) narrates a fine Nova episode that summarizes the Grand Tour of all four Jovian; it is titled, "To Boldly Go", and available via PBS videos. Contact your local PBS station for details on ordering it.

Carl Sagan's Cosmos 106, "Traveller's Tales", only covers Jupiter in detail, but is a good introduction to this chapter. In Cosmos 103, Sagan discusses the possibility of life forms floating in the atmospheres of planets like Jupiter.

Arthur C. Clarke's "2001" brought attention to the Jupiter system, but a far better and more vivid look at Jupiter's satellites is in the sequel, "2010"--encourage your students to view and report on it for extra credit. Both should be available in the Science Fiction section at your local video rental store, as are the Cosmos and Astronomers videos, I imagine.

SUPPLEMENTARY ARTICLES

1. "Looking for Planet X via Gravity," Astronomy, August 1988, page 30.
2. "Pluto and Charon," Astronomy, September 1988, page 52.
3. "The History of the Grand Tour," Astronomy, September 1989, page 44.
4. "Probing Saturn and Titan with Starlight," Astronomy, November 1989, page 50.
5. "Neptune Revealed," Astronomy, December 1989, page 22.
6. "Big and Blue: Uranus and Neptune," Astronomy, October 1990, page 42.
7. "Weather at Neptune," Astronomy, September 1991, page 38.
8. "Saturn, Lord of the Rings," Astronomy, September 1991, page 72.
9. "To the Edge: A Mission to Pluto," Astronomy, May 1992, page 34.
10. "Ulysses Meets Jupiter," Astronomy, July 1992, page 44.
11. "Are there a Thousand Other Plutos?," Astronomy, September 1992, page 40.
12. "The Titan-Triton Connection," Astronomy, April 1993, page 26.
13. "Violent Volcanoes of Io," Astronomy, May 1993, page 40.
14. "Exploring the Solar System on a Budget," Astronomy, August 1993, page 38.
15. "Pluto and Charon," Astronomy, January 1994, page 40.
16. "Jupiter's Smash Hit," Astronomy, November 1994, page 34.
17. "To Boldly Go," Astronomy, December 1994, page 34.
18. "The Mountains of Io," Astronomy, January 1995, page 46.
19. "Hubble Maps Titan's Unseen Surface," Astronomy, February 1995, page 44.
20. "The First New Planet: Discovery of Uranus," Astronomy, March 1995, page 34.
21. "Making Sense of the Great Comet Crash," Astronomy, May 1995, page 48.
22. "The Vanishing Rings of Saturn," Astronomy, June 1995, page 70.
23. "NASA's Discovery Program," Astronomy, November 1995, page 36.
24. "Galileo at Jupiter at Last," Astronomy, January 1996, page 36.
25. "Into the Maelstrom: Galileo Probe Results," Astronomy, April 1996, page 42.
26. "Saturn's Missing Rings," Astronomy, August 1996, page 66.
27. "Neptune's Discovery," Astronomy, September 1996, page 42.
28. "The Lost Ring of Saturn," Astronomy, September 1996, page 22.

ADDITIONAL DISCUSSION TOPICS

Life on Europa?

If the ocean of liquid water Zeilik describes just beneath Europa's fresh ice does exist, then marine life may be possible on Europa, as Clarke brought out in the book version of 2010. If the veneer of ice were thin enough, some sunlight might penetrate; great mats of algae, much like under the ice caps of earth, might flourish just under the ice. Even if the ice is so thick that little sunlight makes it through, the discovery of abundant geothermal life in the deeps of our oceans, and the volcanic eruptions next door on Io, make this form of life another exciting possibility. This was enhanced with the finding of traces of oxygen in the atmosphere of Europa, possibly the result of some type of photosynthesis? When the Galileo orbiter passes closest to Europa in November 1996, its photos and other sensors over a period of several years may show some changes in the water and ice patterns that might indicate biological activity. Ask the students to use the polar caps of earth as a good approximation of Europa's surface, and get them to think of what our satellite data, of similar resolution to the Galileo's cameras, could tell of life under or even on the ice.

Life on Io?

Certainly the violent volcanism might make life on Io seem most unlikely, but remember that liquid sulfur is not nearby as hot as molten rock, and recent Hubble observations have now found water vapor in the thin atmosphere of Io. Geothermal vents, hot springs, et c. come to mind....

Life on Titan?

In Cosmos 106, Sagan speculated about the methane and other hydrocarbons at Titan producing complex organic molecules and even life. In fact, at Titan methane operates in much the same fashion that water does here. It is near its triple point, and exists on Titan in all three states. Methane vapor in the nitrogen atmosphere condenses into droplets and clouds, which cause methane showers to fall on the surface. There lakes of liquid methane (and ethane as well, probably) form. At the colder polar regions, methane probably freezes into methane ice in polar caps. There was some speculation from the Voyager data that Titan might be covered with an ocean of methane kilometers deep, but recent radar studies of Titan at Arecibo have shown a pattern of variable reflectivity, indicating both rough uplands and smoother lakes of liquid. This was confirmed by Hubble Space Telescope data in the February 1995 issue of Astronomy.

But can methane, or any other common liquid, supplant water in its critical role as the fluid of all known life? I doubt it. Water's molecular makeup allows it to form both polar and non-polar chemical bonds easily. Find any other common solvent that can dissolve both a teaspoon of salt and sugar in a glass, and you will have a fluid that can carry both critical forms of nutrients inside a living cell; methane is strictly non-polar, and has no affinity for any salt or ionic compound. At Titan's cold temperatures, water is going to be hard ice, not a life-giving liquid. I would still bet on Europa.

We should have much more data in the next decade. Not only is the Galileo Orbiter to arrive at Jupiter in December 1995, but the Cassini mission to Saturn is being readied for a late 1996 launch. While Galileo's probe is to enter the atmosphere of Jupiter, the probe aboard Cassini will be targeted for Titan, not Saturn. Of course, the orbiters have wonderful cameras and other detectors for studying the satellites close up.

Volcanism at Saturn's Moons?

While the volcanoes of Io are famed, there is a good chance that Cassini will also find active volcanism among Saturn's satellites. Iapetus probably has none now, but the dark material appears to have flowed from underground in places, just flooding the low-lying areas, much like the formation of our lunar mare. The best candidate for presently active volcanism is Enceladus, where one hemisphere is much younger and less cratered than the other. The younger side shows grooved terrain and tectonic features much like those on Europa and Ganymede, and as at Io, there is a ring of debris in the orbit of Enceladus that may have come from active plumes throwing material upward. This "E" ring was unexpected; the collision with a piece within it almost blinded Voyager II in 1981.

Life in the Oceans of Uranus and Neptune?

If the majority of the bulk of the two outer Jovian is in the form of water, and there is abundant heating from the core of each world, then the exciting possibility of life in those very deep, cloud-covered oceans exists. We think of life here being dependent upon sunlight, but remember the deep ocean thermal vent communities get their energy from within the planet, not via photosynthesis. This might be true at Uranus and Neptune, although how we could find out for sure is beyond our present technology. With such huge oceans, I image science fiction writers of the future can visualize some huge monsters for "Sea Quests" to come.

Observational Effects of Comet SL-9's Impacts on Jupiter

"I consider myself lucky to have been a witness to this amazing event. I was able to observe throughout July and August the results of the Jupiter impacts. Using a Panasonic CCD camera attached to a 10" Meade Schmidt-Cassegrain telescope I, and others, attempted to view and videotape the impacts of fragments "B" (on July 16, 1994) and "V"(on July 21). I am sad to say that we were unable to see either collision (both fragments were duds, and no one saw notable plumes from either). These were the only two fragments hitting Jupiter while it was visible in the evening in Pensacola. The following night, while reviewing the footage recorded during the period of the B impact, someone noticed a small black dot on the face of Jupiter. This was the visible result of the A impact about four hours earlier. As more comet sections pummeled Jupiter more black dots arose. On July 22, my seven year old brother Trevor drew Jupiter looking like a smiley face, with two great impacts (D/G and H) forming the eyes, the most obvious features ever seen on the face of another planet in human history. Over a period of days these dots formed smudges, and the smudges began to run together until they had created a new dark gray belt around Jupiter by the end of August 1994. Shortly after this Jupiter was no longer visible to us. On a cool January morning in 1995, at a dreadfully early time, I set foot outside to observe and see if the ring still remained. To my disappointment I found the immense planet to be back to itself again, as if nothing ever happened."

This account of the astronomical event of the century is by Michael Eric Wooten, based on his observations when he was thirteen; his observational project went on to win the Regional Science Fair Grand Award, place fourth in the Florida State Science Fair, and seventh in the Astronomical League's National Junior Astronomer Award in 1996.

When David Levy and Gene and Carolyn Shoemaker happened upon a elongated blur near Jupiter in March 1993, it was the strangest looking comet anyone had ever seen. Larger scopes revealed it to be "a string of pearls," as David Levy put it.

Analysis of the motion of the 21 fragments revealed they were created when the parent comet nucleus passed within Jupiter's Roche Limit, about 30,000 miles above the cloudtops, on July 8, 1992. No one saw this happen, but the pieces exposed much more ice to solar heating, making the comet fragments visible to observers nine months later.

Further orbital analysis by Brian Marsden revealed that Jupiter was pulling the fragments back toward it, and that they would hit near the morning side at about latitude -44 degrees between the 16th and 22nd of July, 1994. All astronomers wondered at what they might see, but hardly anyone dreamed the hits would be as obvious as Michael described. The Hubble and larger earth based scopes showed plumes of hot gases, spurting upward like mushroom clouds from the explosions of many megatons energy. They grew high enough to be seen over the morning side limb, even they the area impacted was still in darkness. The bright plumes lasted only a few minutes before they spread, darkened, and fell back to shower the tops of Jupiter' clouds. Only Galileo, most of the way to Jupiter already, was properly sited to watch the "W" fragment impacting on the night side of Jupiter on July 22, 1994.

Why were the impacted areas so dark? Michael likened it to burned soot, as the heat of the meteors in Jupiter's atmosphere fried many carbon compounds, both in Jupiter's atmosphere and on the cometary nuclei. The amount of darkening were immense, considering that probably none of the fragments was over a mile across, we now think. But then as anyone who has spilled a little ink on a white shirt knows, a little dark stain can go a long way. Apparently the rapid mixing of the convective currents in Jupiter's atmosphere led to the merging of the spots and their gradual fading.

Are events like this really that rare? Jupiter is an immense target, and is known to have trapped several comets with its gravity in the past, generally throwing them back out to the end of the solar system, not eating them, however! It is possible several cases of sudden dark markings on Jupiter may come from similar collisions in the past. We were indeed very lucky to catch this one in time to be prepared to make to most of it!

DEMONSTRATIONS AND OBSERVATIONS

Jovian Globes

We must change our scale here; Jovian globes on the diameter scale we used with the terrestrial bodies will be far too large to use. Also, the lower density makes Styrofoam spheres a good choice, with just a little plaster of paris added to the center. The fact that Jupiter and Saturn are similar in size, as are Uranus and Neptune, means that only two spheres are needed. Try getting a 6" sphere for Jupiter and Saturn, and a 2" sphere for Uranus and Neptune. For Jupiter, the hemisphere should weigh 1.2 kilograms, or about 2.6 pounds. The Saturn sphere should be a little smaller (just sand off a little of the outer Styrofoam, if you like, but only weigh a third as much as your Jupiter globe (Saturn has such a low density it could float, if you had a huge lake to drop it in). The 2" hemispheres for Uranus and Neptune will both need to weigh about 13 grams, yet both will be slightly denser than water. In making these Styrofoam hemispheres, scoop out some foam from the middle, and mix in wet plaster of paris and foam until you get close to the desired weight for the hemisphere. As many paints "craze" Styrofoam, you might like to use colored chalk and a spray fixative to decorate your Jovian models.

For Pluto and the moons of the Jovian, revert back to your terrestrial scale. Pluto will be about 3" across, as large as our moon, but only about 2/3rds as dense, according to the latest findings. Use Styrofoam and enough plaster of paris for a total weight of 220 grams, or just half a pound. In modeling the moons, your moon globe will approximate both Io and Europa; you will need a new, less dense globe for the three largest outer moons. Ganymede, Callisto, and Titan will need to be about 4" in diameter (similar in size to Mercury), but with bulk densities close to 2 grams per cc., only about a third as heavy as the Mercury model; I calculate a weight of about half a kilogram, or just about a pound for a Styrofoam hemisphere with a LOT of plaster of paris added.

For a Pluto-Charon orbital model, use a Styrofoam sphere about 2" across, colored red for Pluto. About 34" away place a 1" Styrofoam sphere, dirty gray in color, to represent Charon. Impale both balls on a rigid 1/4" black wood dowel to remind the students that both bodies rotate in synch-- not only does Charon always face Pluto, but Charon's tides keep the same hemisphere of Pluto always facing Charon.

Observing Jupiter and Family

Of all the planets in the telescope, Jupiter is usually the most interesting. As already discussed, the changing positions of the four Galilean satellites night by night is fascinating even in binoculars. With a 60mm refractor at 30x, the two equatorial belts on Jupiter are visible, as were the largest dark spots during the Comet SL-9 impacts in July 1994. Have your students note the alignment of the prominent equatorial belts with the plane of the moons' orbits. With a 100mm reflector (4") at 100x, you can probably spot the Great Red Spot coming onto the disk every ten hours as Jupiter rotates. The GRS is not always great or even red, however; like just about everything else on Jupiter's turbulent disk, its visibility changes week by week; in fact in 1990, the south equatorial belt vanished for about six months, the first time that has been observed ever. It is always interesting to watch Jupiter coming out into the morning sky from behind the sun just after conjunction, to find out what major changes in the Jovian atmosphere have occurred in the past two months, with Jupiter lost in the sun's glare.

With a six inch reflector and larger, you can see the satellite transits and occultations. Note that Sky and Telescope gives you a monthly listing of such phenomena. The "Dance of the Planets" program will let you visualize them for any day and hour on the computer monitor. Your students will be fascinated to watch the black dot of one of the moon's shadows march across the face of the giant planet, or gasp as they see a moon's disk fade in seconds as it reaches Jupiter's shadow. Remind them that while Jupiter's moons are not that much bigger than our own, Jupiter is much farther from the sun, so the umbral shadow of each moon is correspondingly longer and larger, and the shadows on Jupiter are darker than the moon's shadow on the earth.

But telescopes far larger than your class has access to will be needed to see any detail on the moons of Jupiter; under good seeing conditions, you may resolve them into disks; note that Io is noticeably orange in color, while Europa is bright white.

You might try sketching the changing features in Jupiter's clouds, but do so quickly; the planet's rapid rotation causes the patterns to shift noticeably in just five minutes. Be ready for major changes in the appearance of Jupiter's atmosphere over time. The differential rotation means the same regions of the equatorial belts come around every nine hours, 50 minutes, but at the higher latitudes of the Great Red Spot, rotation takes an additional five minutes.

Amateur astronomers of the Association of Lunar and Planetary Observers have for years monitored the currents of Jupiter's atmosphere and contributed much to our understanding of the dynamics of the meteorology of the giant world.

Observing Saturn

The treat is Saturn's famed rings. They can be resolved with a 60mm refractor at about 40x, and you will probably also notice the large moon Titan, well removed from the ring system, orbiting Saturn in about a month. With a 4" reflector at 100x, you can expect to see a split (Cassini's Division) in the rings at 100x, and probably pick up Tethys and Dione orbiting closer to the planet. With a 6" reflector, more moons of Saturn than of Jupiter can be spotted; Jupiter's fifth largest moon, Amalthea, is smaller than seven of Saturn's moons, and requires an observatory class telescope to spot. You can probably also see two equatorial belts, similar to Jupiter's in location but not as noticeable, and a bluish polar cap near solstices. The weather on Saturn is far less turbulent than at Jupiter, for Saturn gets less heat from the distant sun, and from its less dense interior than Jupiter.

While it is not something you can easily show in one term, it is worth noting that the ring appearance changes dramatically with Saturn's seasons. Saturn's tilt is 27 degrees, similar to our own, while Jupiter's tilt is only 5 degrees, so its equatorial belts always face sunward. This was most noticeable in 1995-6, when Saturn reached an equinox. Then the rings were pointed directly at earth and sun. They are so incredibly flat and thin that they vanished for several weeks with most smaller telescopes. Because the broad expanse of the rings provides a huge reflecting surface, the brightness of Saturn will drop a good deal during this period. Saturn is more than twice as bright at solstices, such as in 2002-3, when the rings are seen most open, than at equinoxes such as in 1995. Tell your students to remember exactly how they observed Saturn, and invite them to come back years later and watch the changing appearances of the ring system. Better yet, have your students make drawings or take photos of Saturn's rings telescopically, and put them on a poster over several years to show the changing tilt as we see it from earth. Be sure to list the student's name and when they took your course!

If you built the 6" sphere model of Saturn mentioned earlier, you can model the rings on the same scale and use your Saturn ring globe, tilted at a 27 degree angle of course, to also show the Saturnian seasons. Use a clear, stiff piece of plastic sheet, cut in a circle about 14 inches in diameter, to stand for the ring. Use figure 11.27 and white paint to draw in the A and B rings and Cassini division if you like.

Now sandwich the rings between the Styrofoam hemispheres of your Saturn globe. Hold the globe and rings up the class. When the rings are most wide open at solstices, they reflect so much extra sunlight that students can see why Saturn's brightness is doubled. Yet at equinoxes they will practically vanish, just as with the equinoxes of 1995-6 and 2010.

Observing the Outer Worlds

For Uranus and Neptune, the best you can hope for is to glimpse their blue-green disks at about 50x or greater with a 60mm refractor or larger. Finding them is best done first with binoculars and finder charts published in January for each year by <u>Sky & Telescope</u> and <u>Astronomy,</u> or in Ottewell's <u>Astronomical Calendar</u>. Note their slow motion through the stars over the course of several weeks, just as Herschel did in 1781 when he chanced upon Uranus.

It is notable that in mid-1993, Uranus passed Neptune in eastern Sagittarius, and for several years afterward both bodies can be glimpsed in the same binocular field with dark skies. This last happened in 1822, 34 years before the discovery of Neptune, so you and your students can observe a historic passage never before seen in human history. It was the fact that Uranus sped up prior to its conjunction with Neptune, then slowed down afterward due to Neptune's gravity braking it, that led Adams and Leverrier to their historic prediction of the placement of Neptune and led Galle to its telescopic discovery in 1846. Read more about the discovery in the September 1996 issues of both Sky and Telescope and Astronomy, which celebrate the 150th anniversary of Neptune's discovery.

Pluto is far fainter than Neptune, being a much smaller body. An eight inch telescope and a very dark observing site are needed to spot it; again finder charts are published in January.

Reproducing Io's Colors

Realize that the color photos issued by Jet Propulsion Lab are overly vivid; the original were taken in black and white, through six different color filters. The final product is far more colorful than you would have seen visually aboard Voyager. Still, if your classroom is a physics room, you probably have a gas burner available. In a flame-proof test tube, gently heat a tablespoon of flowers of sulfur as the students watch. The sulfur melts under low heat, turning into a cherry red fluid looking much like Vick's cough syrup. Continued heating causes the sulfur to grow darker, and flow more slowly, much like molasses. Any more heat will cause it to burn with a blue-green, choking flame. Put this out immediately and use a hood if you can to disperse the fumes, but you might note that the sulfur dioxide formed crystallizes out white as snow. Have your students looks at color photos of Io and relate the colors they noted to the various temperatures on Io's surface.

Dancing with the Jovian

If you are using the IBM PC program, "Dance of the Planets", you can call up some great visualizations of eclipses and planetary phenomena. First go to M (menu), then E (earth view), then F (find). Type in "fix Jupiter", then chose I (invert) to give you telescopic views; Z (zoom) to about 512 X, and pick up the P (pace) to 50 X of so to watch the moons of Jupiter orbit and cast their shadows while the giant planet rotates. Do the same for Saturn, and use D (date) to change years to show the varying tilts of the rings as Saturn passes through its seasons.

ANSWERS TO STUDY EXERCISES

1. Jupiter is far more massive than all the other planets combined, making up about seventy percent of the mass of all the planets.

2. The flow patterns of the belts and zones of Jupiter, particularly around the Great Red Spot and some of the white ovals. The Red Spot itself would be subject to continued observation.

The Great Red Spot is not always so great or red, either. In the 1990's, twice the South Equatorial Belt has faded greatly, unprecedented in previous telescopic observations of the giant world. The rapidly moving shadows of the four Galilean moons would also be fascinating as they move across the belts and zones.

3. From recent eclipses of Pluto and Charon, we have found Pluto to have a diameter of about 2,300 km. From Charon's orbital period of 6.4 days and distance of 19,100 km, we get Pluto's mass as about a fifth of that of the moon, or about 1.4×10^{22} kg. Thus Pluto's density is 1.44×10^{22} kg / $1.333 \times 3.14 \times (1.15 \times 10^6$ m$)^3 = 2,260$ kg/m^3, about two thirds the moon's density. Like the moon, Pluto is fairly rocky, but with more ice close to its surface.

4. When Saturn is at its equinox, as in 1994-95, the rings vanish, being too thin to be visible in even the largest earth-based telescopes (maybe Hubble will do better); they can't be more than a few kilometers thick. Also, then the rings are most open at solstice (next in 2002), we can see stars through the rings. Also, Voyager I and II both showed the rings were very thin.

5. The Jovian planets have many moons, and any of them can be used with Newton's Revision of Kepler's Third Law to weigh the Jovian they circle.

6. The Jovian are much less dense, being made of lighter hydrogen and helium, gases the terrestrials lost long ago.

7. The ring system of Saturn is much larger and more extensive than the sparse and dark rings surrounding the other three Jovian. Saturn's ices are highly reflective and thus bright, while the rings of Jupiter are made of rock dust, while the dark rings of Uranus and Neptune seem to be coated with carbon dust.

8. Jupiter's magnetic field is a dipole field inclined slightly (about 11 degrees) to the planet's rotation axis, much like earth's. However, it is far stronger than the earth's, and probably generated by a liquid metallic hydrogen mantle rapidly spinning, unlike the molten iron-nickel core of earth. Saturn's field is even less tilted than Jupiter's with respect to its rotation, but otherwise quite similar.

9. Pluto is slightly smaller than Io or Europa, and a little less dense than both, indicating a higher ice and lesser rock content than for the two inner Jovian moons; however, its density is greater than that of either Ganymede or Callisto.

10. Except for rocky Io and Europa, most Jovian moons are made chiefly of water and other ices, with some rocky material chiefly near their cores.

11. The more craters, the older the surface.

12. Uranus' magnetic field is similar to the earth's in strength, suggesting that iron rather than metallic hydrogen generates it. Even more peculiar, the field is tilted vastly off axis. It is nearly sixty degrees off and the magnetic field does not even pass through the core of its planet. Figure this out and take another trip to Stockholm.

13. While Neptune is only tilted 29 degrees off the ecliptic, and Uranus' tilt is 98 degrees, both magnetic fields are tilted about 60° off their rotation axis, and neither passes through the planet's core.

14. The earth's core is molten and solid metals, chiefly iron and nickel. Jupiter may have a similar metallic core, but surrounding it is a metallic hydrogen mantle, a phase that earth's gravity can not replicate.

15. Charon is half as large as Pluto, and its tidal pull is great enough to keep the same face of Pluto always facing its moon. Both rotate and revolve around their common center of gravity every 6.4 earth days.

ANSWERS TO PROBLEMS & ACTIVITIES

1. $g = GM/R^2 = (6.67 \times 10^{-11})(1.9 \times 10^{27})/(7.14 \times 10^7)^2 = 25$ m/s^2, about 2.5x greater than earth's.

2. $M_{pluto} + M_{charon} = 4 \times (3.14)^2 \times (1.75 \times 10^7 \text{ m})^3 / 6.67 \times 10^{11} \times (5.5 \times 105 \text{ sec})^2$, giving a mass of about 1.1×10^{22} kg, about one fifth the mass of the moon. You also need to know where the center of mass for the two body system lies, in other to divide this total mass between the two bodies.

3. Using Io data, $M_{jupiter} = 39.5 \times (4.22 \times 10^8 \text{ m})^3 / (6.67 \times 10^{-11})(1.53 \times 10^5)^2 = 1.9 \times 10^{27}$ kg.

4. For an angular size of 47 arc sec., we get a size to distance ratio of sin .013 = .000228 or Jupiter is 4,388 x its diameter in distance. Multiplying this times its equatorial diameter of 142,800 km, we get its distance as 627 million km. The Pluto reference is incorrect; the quoted 3 arcsec is the diameter of NEPTUNE at closest approach to earth. This gives a size to distance ratio of 68,754, which multiplied by Neptune's diameter of 49,600 km gives us a distance of 3.4 billion km to Neptune. The closest approaches happen at opposition for both bodies.

5. Jupiter's escape speed of 61 km/sec is about 5.5x greater than earth's, explaining why Jupiter can hold on to light hydrogen and helium which the terrestrials lost long ago. But Pluto's escape speed is $(2 \times 6.67 \times 10^{-11} \times 1.1 \times 10^{22} \text{ kg}/1.15 \times 10^6)^{.5} = 1.12$ km/sec, about a tenth the earth's.

6. Neptune gives off 3x as much heat as it gets from the sun. At Neptune's distance of about 30 A.U., this means by the inverse square law that Neptune only gets 1/900th the solar flux at earth, where the solar constant is 1,370 watts/square meter. Thus neptune is radiating about 4.5 watts per square meter; given its radius of 24,800 km, we get a total surface area of the sphere as $A = 4 \times 3.14 \times (2.48 \times 10^7 \text{ m})^2 = 7.72 \times 10^{15}$ m^2; multiplying this by the 4.5 watts per square meter, we get a total radiation of 3.5×10^{13} kilowatts.

CHAPTER 12

The Origin and Evolution of the Solar System

CHAPTER OUTLINE

Central Question: What physical processes resulted in the formation of the planets from a cloud of gas and dust?

CHAPTER OVERVIEW

Teacher's Notes

Unlike the preceding chapters, this chapter contains a bit of factual material and much speculation. You should emphasize just how uncertain that details are in the scenario for the formation of the solar system. This point is a good example of the process of model building, and of incorporating new ideas into existing models to "patch up" difficult spots.

To illustrate both strengths and problems with the nebular models, try making up a table listing dynamical and chemical properties of the solar system, and then show how well each model meets these requirements. To extend the comparison, have students propose ways to clear deficiencies in the models, emphasizing that new proposals must "fit in" to the general scheme. For instance, could the angular momentum problem in the nebular model be rectified if the planets formed before the sun did, and so retained the angular momentum of the nebula before it became centrally concentrated? Would this interfere with the condensation sequence and introduce problems with the observed chemical distribution within the solar system?

One dynamical characteristic not emphasized in this text is that planetary distances from the sun follow a regular spacing rule, the so-called Titus Bode Law. While no dynamical or physical basis for Bode's sequence has been found, it does serve as a useful way to remember planetary distances (at least out to Uranus). Its success in predicting most planetary distances is remarkable and suggests that a physical explanation may yet be found; there is still scientific interest in finding that basis.

Some additional discussion topics include the detection of planetary systems around other stars. Nebular models predict that the number of planetary systems should be large. A few stars in the solar neighborhood have irregular motions that suggest orbiting bodies around them; Barnard's stars is one of the most notable and controversial examples. Even some pulsars seem to have planetary mass objects close to them, based on Doppler shifting of their radio pulses! Also, the infrared observations from space and high mountaintops have revealed nebular disks around several stars, such as Vega, Fomalhaut, and most notably beta Pictoris. Note that these disks show up better in the infrared, the visible light absorbed by the dusty disks, to be emitted as heat by the dust grains.

12.1 Debris between the Planets: Asteroids

The solar system, including the sun and planets, grew from the accretion of small debris. By studying the surviving smaller bodies, such as comets, asteroids, and meteorites, we may gain much insight into the early history of the solar system.

Most asteroids are not now differentiated bodies; Zeilik notes that their brightness varies as they rotate, probably due to irregular surfaces and shapes. The Galileo photo of Gaspra on page 249 looks quite similar to the photos of Deimos and Phobos and to the new photos by Galileo of another asteroid, Ida, returned to earth in October 1993. But the parent body of at least some asteroids must have been differentiated; note that M-type bodies have metallic reflectivities much like the core of a broken up parent body that had time to differentiate.

The S-type asteroids are the most abundant, and made of stony materials like our mantle (which makes up 5/6ths of the earth's volume), and the C-type have carbon rich coverings, typical of the black surfaces we saw in the outer solar system on Iapetus, and in the rings of Uranus and Neptune. Note the density gradient; the C-type are lowest in density of the three, and farther out.

Why do we find thousands of smaller bodies between Mars and Jupiter, instead of a terrestrial planet? The debris we now find in the belt, if lumped together, would accrete into a body smaller and less massive than our own moon, but all these are probably just the surviving remnant of a once much larger population.

Part of the answer lies in the condensation sequence; there are the number of high density metal compounds that are plentiful and condense out between 1500 K and 680 K; it is this material which condensed close to the sun that formed the terrestrial bodies. Then there is water ice and other ices, all condensing out at less than 175 K; these would accrete into the gas giants and their icy families. But between there is no really common material to condense out in the region today occupied by the asteroid belt; no massive planet could form in this sparse region.

Another cause is that Jupiter's great gravity perturbed the orbits of the larger asteroids, causing them to collide with each other so violently that both bodies shattered. Had the collisions been less energetic, the bodies would have accreted into larger planets, instead to leaving behind the shattered debris we now observe. Even collisions with larger outer moons of Jupiter might have shattered any world pulled outward by Jupiter, with the heavier stony and iron-rich materials pulled inward toward the sun, while the low density ices would have been flung far afield, to contribute to the cometary cloud. The violence of these collisions probably threw most of the material out of the main belt. With orbits that now crossed the planet orbits, these bodies were quickly swept up and produced the cratering we observe on the surfaces of the solid bodies throughout the solar system. It is worth noting that the cratering record reveals sudden infusions of new craters about 600 million years after the solar system's formation--this may date to the break up of one or more of the parent bodies of the asteroids.

12.2   Comets: Snowballs in Space

A comet differs from a C-type asteroid chiefly in having a higher fraction of ices and other volatiles. If you were to have a comet pass frequently close to the sun, and loose these volatiles, it might completely disintegrate, or turn into a dark, earthgrazing asteroid like Phaeton, the parent body to the Geminid meteor shower. It was taken to be such as asteroid when discovered in 1983 by the IRAS (Infrared Astronomy Satellite); later analysis of its orbit showed its cometary connection. Just as Pluto lies between planets and comets, so Phaeton lies between comets and asteroids.

Halley's is not the biggest or brightest comet, but it is the only short period comet for which we have a good track record. It is big enough that it has been seen visually at every apparition since 239 BC. The 1986 return could not have been worse; perihelion occurred with the comet on the far side of the sun, hopelessly lost in the sun's glare. Still, the comet was visible from late November 1985 until mid-January 1986 with the naked eye in the evening sky. It returned to visibility, now on the morning side of the sun, in late February 1986, and headed straight south, so the northern U.S. and Europe could not see it below their southern horizons in early April when it was closest to earth.

Even those of us who did head south were disappointed, for the comet shed its gas tail in late March and the dust tail, while bright, was short and fan shaped, not the long flowing appendage seen so well in 1910. But in 1910, the comet came three times closer to earth, was nine times brighter, and we even passed through the gas tail (without incident) in May 1910. The 2061 return will be far better for northern hemisphere viewers, so inspire your students to live long enough to see it return. By then we will have probes to actually land on the nucleus of this famous visitor from out there.

Note how much the dark nucleus of Halley's resembles Gaspra and Phobos. Again, except for forming farther from the sun, and gaining a richer supply of volatiles, C-type asteroids and comets are generally composed of similar materials. Chiron, a large asteroid orbiting between Saturn and Uranus, was recently seen to have a coma, and be reclassified. The December 1992 Astronomy contained a brief report on a comet which was active at the last solar encounter in the 1940's but showed no activity when recovered in late 1992.

12.3 Meteors and Meteorites

As a student pointed out to me, a meteoroid is "in the void"; the only thing that distinguishes it from an asteroid is that it is too tiny to be yet seen. The meteor actually last only seconds; most burn up entirely, to reach earth only as a fine shower of dust (in many cases the seed particles for rain showers--all of us have been pelted by microscopic meteor showers in the past). The few which do survive are elevated to meteorite status if they make it to the surface almost intact. Practically all meteorites are made of denser, more substantial asteroidal materials. Most cometary debris is too fragile and burns up in the atmosphere. Nor can it be seen coming; the press made a big deal over the "Perseid Storm" in August 1993; it even delayed the launch of a Space Shuttle mission. Many of the public were upset when the storm did not materialize, but this fine debris can't be detected until it hits our atmosphere to glow as meteors.

Have the students match the meteorite types to their likely site of origin in the parent asteroid. Irons must have come from the core, stony-irons from the core-mantle boundary, stones from the mantle, and carbonaceous chondrites from the dark crust, much like that of Deimos, Phobos, and Gaspra. Note that the irons and stones have come deep within the parent body, and been subjected to considerable heating. By contrast, the carbonaceous chondrites contain volatiles and organic compounds easily decomposed by such heating; they have been kept cool throughout their history. Most agree with Zeilik that these meteorites are as close as we can get to the basic stardust from which our solar system condensed.

12.4 Pieces and Puzzles of the Solar System

The regular spacing of the planets referred to on page 258 is the Titus-Bode Rule. To get the distances to the planets in A.U., start the sequence 0,3,6,12,24, etc., then add 4 to each number (why? It works...) and divide by ten. The sequence is then .4, .7, 1.0, 1.6, 2.8, 5.2, 10.0, and 19.6. It closely matches the distances to the planets from Mercury out to Uranus, leaving room at 2.8 A.U. for the asteroid belt. It was the prediction that a planet should lie between Mars and Jupiter that led Bode to organize the "Celestial Police" who would discover the four largest minor planets. Perhaps a massive body perturbed the orbits of Neptune and Pluto, for they do not fit the pattern well at all; in fact, Pluto's average distance of 40 AU is a fair fit for the Titius-Bode prediction of the distance to Neptune. It does not presently appear this pattern is being followed by other solar type stars, however. Of the seven new solar systems found in 1996, most have a large Jovian type planet relatively close to their star, in the region we find the terrestrial planets around our sun. For a look at the variety of solar systems found to date, see "Other Stars, Other Planets" in the August 1996 issue of Sky and Telescope on page 20.

12.5   Basics of Nebular Models

Recent research has found a nebular cloud already in a disk, probably with terrestrial type planets condensed out of it, around the star beta Pictoris. Many other nearby young stars, including Vega and Fomalhaut, seem to have dusty disks, strongly supporting the nebular theory and the idea that planetary systems are abundant in the galaxy.

Also supporting the transfer of angular momentum are observations that the older stars in the Pleiades cluster are spinning slower than the youngest ones. If this is the case, then the magnetic fields of the older stars have already transferred enough energy outward to create the formation of disks around them, with planets soon to follow.

12.6   The Formation of the Planets

Consider condensation similar to the formation of snowflakes, accretion more like violent updrafts rolling them together to produce hailstones. Materials of higher melting and boiling points would be left close to the sun, while more volatile materials would be driven to the outer regions of the nebula to form the jovians and their moons.

ADDITIONAL RESOURCES

Slide 12.15 (newly recovered Comet Swift-Tuttle) and transparencies 12.1, 12.5, 12.12, 12.18abc, 12.19, 12.21ab, 12.22abcd, and 12.23 will update your class presentations.

The "Tales from Other Worlds" video from the Planet Earth series fits well with this chapter, if you did not already use it with the terrestrial planets. The PBS video, "Comet Halley--Once in a Lifetime" is somewhat dated, but does capture well the unpredictability of comets. The Nova episode on Asteroids encountering early in the summer of 1996 is very well done.

The video "Meteor," starring Sean Connery, is widely available at rental stores. It looks at a possible collision between earth and an earthgrazing asteroid. Have your students view it after they have read Brian Marsden's article on Comet Swift-Tuttle in the January 1993 issue of Sky and Telescope, and critique how accurate they think the video was.

If you can, get your school to order a small meteorite collection. Check Earth or some current astronomy magazines for listings of dealers; write off to each for prices and listings of specimens available. Try to get at least a good stone and an iron; the students will be amazed at the differences in density of these pieces of the asteroid belt. If your iron has been cut and polished (a good idea, and well worth the extra price), show the students the Widmanstatten figures in the iron. These large crystal patterns tell us the irons formed deep inside the parent body, and cooled very slowly, forming the large iron crystals. No iron bearing mineral at or near the earth's crust has such a pattern as these. Remind the students they are holding the heart of a planet, now broken apart, in their hands. Better watch your security, however. We lost several nice meteorites from a display case at PJC in 1995; good meteorites are worth hundreds of dollars.

NASA's Lunar Receiving Lab also maintains clear disks with meteorite specimens, which are visually more interesting than the rather dull looking moon rocks discussed back in chapter nine. Try to borrow their meteorite disk; it comes with a nice slide set and taped narration, and a good deal of information on the particular pieces in your disk.

SUPPLEMENTARY ARTICLES

1. "My Most Memorable Comets," <u>Astronomy,</u> August 1986, page 24.
2. "What We Have Learned About Comet Halley," <u>Astronomy,</u> September 1986, page 6.
3. "Halley's Nucleus--Search for the Primitive," <u>Astronomy,</u> June 1987, page 96.
4. "How to Hunt for the Elusive Meteorite," <u>Astronomy,</u> August 1989, page 70.
5. "Chiron: Asteroid or Comet?," <u>Astronomy,</u> August 1990, page 44.
6. "Where do Comets Come From?," <u>Astronomy,</u> September 1990, page 28.
7. "Voyages to Worlds of Ice," <u>Astronomy,</u> December 1990, page 42.
8. "Fireballs from Outer Space," <u>Astronomy,</u> February 1991, page 64.
9. "Demise of the Dinosaurs," <u>Astronomy,</u> August 1991, page 30.
10. "Asteroid Impact," <u>Astronomy,</u> September 1991, page 50.
11. "Will the Lion Roar Again?," <u>Astronomy,</u> November 1991, page 44.
12. "Galileo Views Gaspra," <u>Astronomy,</u> February 1992, page 52.
13. "Mysterious Sungrazers," <u>Astronomy,</u> April 1992, page 46.
14. "Captive Asteroids," <u>Astronomy,</u> July 1992, page 40.
15. "Interplanetary Fugitives," <u>Astronomy.</u> August 1992, page 30.
16. "Are there a Thousand Other Plutos?," <u>Astronomy,</u> September 1992, page 40.
17. "Comets' Role in the Origin of Life," <u>Astronomy,</u> November 1992, page 28.
18. "When Asteroids Collide," <u>Astronomy,</u> February 1993, page 36.
19. "Motes in the Solar System's Eye," <u>Astronomy,</u> May 1993, page 34.
20. "Spuds in Space," <u>Astronomy,</u> July 1993, page 34.
21. "The 1993 Perseid Meteor Shower," <u>Astronomy,</u> August 1993, page 76.
22. "Jove's Hammer--Comet 1993e hits Jupiter," <u>Astronomy,</u> October 1993, page 38.
23. "Giotto's Second Cometary Encounter," <u>Astronomy,</u> November 1993, page 42.
24. "Death from the Sky," <u>Astronomy,</u> December 1993, page 38.
25. "Catch the Geminid Meteor Shower," <u>Astronomy,</u> December 1993, page 82.
26. "Chiron: Interloper from the Kuiper Belt," <u>Astronomy,</u> August 1994, page 26.
27. "To Boldly Go," <u>Astronomy,</u> December 1994, page 34.
28. "Double Asteroids Mean Double Trouble," <u>Astronomy,</u> January 1995, page 30.
29. "Seeing the Zodiacal Light," <u>Astronomy,</u> March 1995, page 70.
30. "Target Earth," <u>Astronomy,</u> October 1995, page 34.
31. "NASA's Discovery Program," <u>Astronomy,</u> November 1995, page 36.
32. "Here Comes Comet Hale-Bopp," <u>Astronomy,</u> February 1996, page 68.
33. "Far Journey to a Near Asteroid," <u>Astronomy,</u> March 1996, page 42.
34. "The Day the Dinosaurs Died," <u>Astronomy,</u> April 1996, page 34.
35. "Two New Solar Systems," <u>Astronomy,</u> April 1996, page 50.
36. "Worlds Between Worlds," <u>Astronomy,</u> June 1996, page 46.
37. "Hooray for Comet Hyakutake," <u>Astronomy,</u> June 1996, page 76.
38. "Showtime for Comet Hale-Bopp," <u>Astronomy,</u> July 1996, page 69.
39. "Hyakutake's Spring Surprise," <u>Astronomy,</u> July 1996, page 74.
40. Catch a Perseid on Film," <u>Astronomy,</u> August 1996, page 52.

ADDITIONAL DISCUSSION TOPICS

Sungrazers

Part of the mystique of comets is their unpredictability. Comets are small bodies, easily perturbed by the gravities of the planets. For instance, in May 1983, our earth threw Comet IRAS-Ariki-Alcock completely out of the solar system in the hyperbolic orbit, just as Jupiter sent the Voyagers out to the stars. The most memorable comets are likely to be the spectacular sungrazers, who come so close to the sun that they loose great quantities of gas and dust and form very bright, easily visible tails. Comet Ikeya-Seki in 1965 was perhaps the best of this century. I remember getting up to observe it on November 1, 1965, and marveling as the tail stretched half way from the horizon to the zenith. It was so bright at perihelion that some observers blocked out the sun and saw it with the naked eye in broad daylight.

But at least eight comets of the nineteenth century outshone Ikeya-Seki, according to historical records; our lives have been lived in a relative dearth of great comets. But many such sungrazers escape detection entirely, for they may come from the direction of the sun, and be entirely destroyed by solar heating at perihelion.

Many such comets went undetected, but in the last decade NASA satellites have spotted several of these. One comet a century ago lost its nucleus but appeared for several evenings in the northwestern evening sky as a ghost comet, with only a tail, its head decapitated by the sun. Even if the comet is not completely destroyed, the fragile icy nucleus may fracture. Comet West put on such a good show in March 1986 because it broke up into four fragments at perihelion. These four pieces exposed far more surface area to solar heating, so it formed a much better tail that expected from its performance prior to perihelion. But the next time it returns, we will see four different smaller comets, all in the same orbit, put following along after each other but separated by several days or even weeks. Examples of such double comets have already been observed.

The Comets of Doom?

As discussed earlier in the chapter on the earth, comet and asteroid impacts have played a major role in establishing the pattern evolution has steered for life on earth. One comet received a great deal of publicity in late 1992 for its possible role in the future of life on earth, while another gave us a close call in March of 1996..

Comet Swift-Tuttle was observed easily with the naked eye in 1862, during the American Civil War. Studies of its orbit led to the expectation that its period was about 120 years, so we expected it back in 1981. In the meantime, studies of the orbits of meteor showers proved that it was the parent comet to the best annual meteor display, the Perseid Meteor shower which peaks in mid-August. A very interesting historical account of the discovery of the Perseid meteor shower is given in the August 1996 issue of Sky and Telescope on page 68. The fact that its debris struck earth showed that this comet's orbit obviously lay close to our own, perhaps too close as thing turn out.

We looked patiently, but no one observed the comet during the 1980's. Some speculated it had disintegrated, but Brian Marsden of Harvard considered that perhaps jets from the comet's nucleus were acting as boosters to speed up the comet into a larger orbit with a longer period. His study of historical records turned up a similar orbit for a comet seen in China in 1737, so Marsden boldly predicted that Swift-Tuttle would return in the summer of 1992.

117

Again no one found it, and hopes dimmed, until a Japanese amateur found it in the handle of big dipper in late September 1992. It grew bright enough to see faintly with the naked eye in late fall of 1992, but the most interesting development Marsden noted is what may happen at its next apparition. If it continues picking up speed, its orbit may grow from the .94 AU perihelion of 1992 to exactly cross our orbit in its return about 2126. If its perihelion passage were to happen about August 12, 2126, then it would collide with the earth, as graphically illustrated in Marsden's article in the January 1993 issue of Sky & Telescope. There is only a 1 in 10,000 chance this will happen, and by then we should have the technology to find ways to be sure the comet misses us. For some ideas, view "Meteor" as discussed above. Later observations as the comet neared perihelion allayed our fears, and we now know the comet will pass the earth's orbit several weeks ahead of us next time around, but we still can anticipate a great meteor swarm from this close encounter at the next comet-earth encounter. Meanwhile, the Perseid meteor shower may be much richer than normal for the rest of the twentieth century, thanks to the comet's proximity.

Comet Swift-Tuttle is another reminder that the behavior of comets is far less predictable than that of planets. Comets are constantly loosing mass to the sun, and their jets can act as retrorockets to slow them down into shorter periods, or boosters to speed them up into still longer periods, as was the case for this comet. It is certainly in our best interest to keep track of such comets and their changeable orbits--this information could someday save all civilization.

Just how things can change was dramatically illustrated early in 1996. A Japanese amateur, Yuji Hyakutake, found two comets within six weeks of each other in late 1995 and early 1996. The one found on January 31, 1996, Comet Hyakutake 1996B2, would soon become world famous.

A note here on the new numbering system to identify comets. New discoveries, as in the past, are assigned the name (or names, such as Alan Hale and Tim Bopp) of the first (up to three) independent discoverers. Instead of just going down the alphabet after the year, as previously done (we were routinely running out of letters), we now assign a letter to each half month of the new year, then assign numbers to indicate the order of discovery within that period. Thus B2 was the second comet found in the last two weeks of January.

What made this comet special was that on March 25, 1996, it passed just ten million miles above our north pole. Billions of people watched the comet, its tail growing longer each night, approach the earth in mid-March, then sweep across the sky from NE to NW in just a week. From dark observing sites, the faint tail could be traced about 60 degrees across the sky, from Polaris to Spica, with the naked eye! Because it was so close to us, the comet was a easy naked eye object, and its blue-green color was intense enough to note visually and photograph well with just 30 second, unguided exposures and fast film. The comet would get much closer to the sun by perihelion in early May, but it was getting lost in the sun's glare and later was too far south for most northern observers to follow it outbound from the sun. Remember, no one new about this comet until January 31, 1996, yet in two months it would be making headlines. Certainly the fame Yuji Hyakutake gained from his discovery will inspire many others to take up comet hunting as well.

In May 1996, we got an even closer call from a small asteroid, 1996 JA, which passed within 270,000 miles of earth, just outside the moon's orbit. While Hyakutake will not be back for thousands of years, the earthgrazing asteroid and many more like him never lie more than a few years away! It is important that projects like Skywatch, conducted by David Levy and the Shoemakers of Comet SL-9 fame, be continued to spot such potential threats to human existence.

118

Does the Oort Cloud Exist?

Oort's original proposal in the 1950's was for a spherical cloud, extending perhaps over a light year out from the sun, or almost a third of the way to alpha Centauri. Such a spherical cloud, so far from the sun, would be susceptible to the gravitational perturbation of any star passing within a few light years of the sun, as alpha Centauri is now doing. Some comet would be stripped away by the star's gravity, and lost in space. Were we to observe a comet inbound toward the sun with such an orbit already, it must be coming from another solar system.

One such "hypercomet" may have been observed in 1989; if such a comet came sunward from beyond the Oort Cloud, it must have been formed around another star--what an interesting specimen a sample from it might be.

But if some comets sped up and escaped, others were slowed down and pulled in sunward on a trip taking hundreds of thousands of years. These are the long period comets, in very elliptical orbits taking hundreds of thousands of years; they would come in random directions, so observers are not able to concentrate their searches for comets along the ecliptic, the way we spot the planets visually. Most swept past the sun once, then were hurled by its gravity back out to the cloud to resume their deep sleep. But a few encountered planets, to strike them as perhaps happened at Tunguska in 1908, or be sped up and expelled from the solar system, or be slowed down into a shorter period. For instance, Halley's can't have been loosing the huge amounts of water vapor and other volatiles long, or it would have already turned into a black cinder like Phaeton. It is possible we have no observations of Halley's earlier than 239 BC because that marks the passage at which Halley's was trapped into its present 76 year period.

In the 1980's some speculated that while long period comets must comet from the Outer Oort Cloud, there might also be an inner belt, better aligned with the ecliptic, of debris expelled by the planets early on. This is also called the Kuiper Belt, and would extend from just beyond the orbit of Neptune perhaps several hundred A.U. outward, but close to the ecliptic plane. Pluto would then be the largest of the Kuiper Belt comets in this new model. In December 1992 came news of the discovery of a cometary nuclei, perhaps 40 km across, in this Kuiper belt region. As of this writing three years later, 30 more of these bodies, several hundred miles across and orbiting between 33 and 38 A.U. from the sun, have been recovered. Stay tuned for more developments as the search continues. Also, the Hubble Space Telescope photographed much smaller comet nuclei, similar in size to Halley and Hyakutake, moving at exactly the speeds predicted for Kuiper belt bodies. They were very abundant, but most have stable orbits that keep them well out of the sun's vicinity.

One body that did blunder in is Comet Hale-Bopp. It is possibly the biggest nucleus on record to enter the inner solar system, and at this writing is 10,000X brighter than Comet Halley was at the same distance from the sun in 1985. In August 1996, Comet Hale-Bopp is already a naked eye comet, and will probably remain so throughout all of 1997. It is traveling northward now, and will be at its best in the northwestern sky in the Spring of 1997. Some estimate this giant comet, often compared to the Great Comet of 1811, will be visible in broad daylight, but it should be far better than Halley's in 1986 in any case. Skymap has the data for Comet Hale-Bopp in its comet.sky file, and you can use it to produce finder charts of the comet in the next few months. It should be a great sight for observers under dark skies, so try to get your classes out to view what may be the very best comet of their lives. Give them enough great memories like this, and you will have no trouble recruiting astronomy students in the future.

Comets, Bringers of Life?

Back in 1985, the famous British astronomer Fred Hoyle made the highly controversial assertion that life originated upon cometary nuclei, and was showered down on the primitive earth like manna from above. While most scientists doubt organic molecules could progress that far on the cold surface of a cometary nuclei, most agree that impacts of comets played a major role in showering the early earth with complex organic molecules, like those in carbonaceous chondrite meteorites, and extra water. Some even think that microcomets striking earth now are still a major supplier of water and organics to the earth. A new science fiction novel, <u>Mining the Oort Cloud</u>, even suggests intentionally sending comets on collision courses with Mars, to both warm it up and supply it with the volatiles the Red Planet has been losing into space. Recommend it to students for extra credit.

Like volcanoes, comet impacts can be both the givers and takers of life. Yin and Yang and all that jazz.

Exploring Near Earth Asteroids

Appealing as those gold mines of Mercury may appear, the biggest need for space colonists of the 21st century will be adequate water. It will be extremely expensive to ship it up from our own oceans, yet it is the vital raw material not found on the Moon and rare on Mars. But if we could mine some of the earth-grazing asteroids, we probably could find great quantities of volatiles like water trapped in the dark carbonaceous chondrite type material. When exposed to solar heating, these could release a good deal of steam and make colonizing the moon, or even building space cities of lunar materials, much easier. Also, the low surface gravity and near earth approaches of several minor planets makes exploring them less of a challenge than exploring the moon via Apollo. Challenge your students to think of the potential for humanity tapping these resources could create. Also read the article about the NEAR mission to Eros in the March 1996 issue of Astronomy.

DEMONSTRATIONS AND OBSERVATIONS

To Catch a Comet

For city dwellers, as those who remember infamous Comet Kohoutek of January 1974 will attest, comets are elusive indeed. They rarely get bright (naked eye ones come perhaps every five years or so) and even then subtle details in the tail are hopelessly lost in the sky pollution of our urban areas. So if you want the fame and fortune that comes with discovering a comet, move to the country and get some really dark skies.

Start by learning the sky very well with binoculars; in fact, several comets are found every year with nothing more than large binoculars. Remember that there are many things that look like comets up there. Messier's catalog is nothing but a listing of the 110 things which he ran across that he mistook at first for comets, so check out your star atlas and be sure you are not observing a previously known cluster, nebula, or galaxy. If that diffuse, bluish blur is not plotted, wait patiently for some motion. To become visible in smaller telescopes, comets must be rather close to the sun, probably at least inside the asteroid belt. In that case, in an hour or so you should be able to detect in which direction and how fast the comet is moving. This information is vital, for now you need to telephone or telegraph the International Astronomical Union at Cambridge, Massachusetts with the location, brightness, and speed and direction of motion of your suspect.

With data in hand, call the Central Bureau for Astronomical Telegrams at (617) 495-7244. If it is confirmed by independent observers (usually within a day--Brian Marsden has a worldwide team he calls up for this), the comet will be assigned your last name, the year of discovery, and the order within that year indicated with a small letter suffix. Comets found in the first half of January are assigned letter A, then numbered in order of their discovery in that time period. As you note from the names, such as Comet Hale-Bopp 1995O1, often two or more comet searchers independently find the same comet almost simultaneously; the names are then hyphenated, with the first to report listed first; only the first three discoverers are so honored.

Even if you do not have the time to spend searching for new comets (typically about 400 hours per discovery, and a lot of luck to boot), you can still enjoy them. While the next good periodic to return is Halley's in 2061 (should be much better than the 1986 apparition), there is always the chance a new comet may round the sun and appear unannounced to millions at the same time one morning, as happened with the Great Comet of January 1910; it had no other name, for millions spotted it as they got up for morning cores that cold winter's morning.

Stephen Smith publishes the <u>Shallow Sky Bulletin</u>, a bimonthly newsletter of new cometary discoveries that can be observed by amateurs. It only costs $10 a year, and keeps you at the "cutting edge" of cometary research. Send your check, payable to the Comet Rapid Announcement Service (CRAS), to CRAS, PO Box 110282, Cleveland, Ohio 44111-0282. The bimonthly finder charts and comet ephemeris may be copied for classroom or astronomy club usage with only a source citation.

In observing comets telescopically, try a variety of magnifications. Low power will reveal the tail and coma best, while higher power views of the region around the nucleus may reveal sudden changes in the jets of gas evaporating off the nucleus. You may even be lucky enough to watch a nucleus split, as happened with Comet West in March 1986. Even if you never see something quite so dramatic, try making sketches of the jets appearance night by night, so see the nucleus spinning and find its period of rotation.

As you use cometary ephemeris to find them with your setting circles or star charts, remember the warnings about cometary unpredictability. Cometary nuclei may be frozen harder than we expected, explaining why Comet Kohoutek was a 100x fainter than was predicted back in late 1973, or more fragile than expected, accounting for Comet West's better showing in March 1976. As the jets coming off comets can act as retrorockets to slow comets down, or boosters to speed them up, even the early predictions of the comet's position can be off by a degree or more; compare the comet's actual position with the prediction, and report any discrepancies to the International Astronomical Union for analysis.

Of Comets and Meteors

Meteors and comets are often confused in the popular press, and even in Hollywood. Look at the opening sequence to "The Butcher's Wife", for instance. The object depicted on the screen is a two-tailed comet, much like Comet West in March 1986. But a comet is millions of miles distant, in interplanetary space. It might move a few degrees per day, but not go flying across the sky as seen in the movie. An object moving that fast is obviously a meteor, streaking through our upper atmosphere. Comets are typically visible for several weeks or months, meteors for a few seconds. As comets are distant, observers on earth, as first noted by Tycho Brahe about 1570, all see the comet against the same background of stars.

121

But meteors are occurring typically 50-100 miles high, so different observers will see the meteor moving in different directions and against a different stellar pattern. In fact, one reason astronomers like to collect fireball reports from many observers over an area of hundred of miles is that by triangulation, we can plot the exact trajectory of the body, both before it struck our atmosphere, and even tell where a surviving meteorite might have fallen.

Observing Meteor Showers

Comets and meteors are connected, however. While most meteorites are fragments of asteroids, most meteors really are cometary debris, too fragile to make it intact to the earth's surface. As an aging comet disintegrates, the debris it leaves behind becomes scattered in the old comet's orbit. If we are fortunate (or, in 2126, unfortunate?) enough for the comet's orbit to intersect the earth's, then this debris can produce a spectacular meteor shower when it strikes our atmosphere. This debris is quite tiny and fragile, for no part of a comet has ever been seen to survive intact as a meteorite striking our surface.

As the earth returns to the same intersection point with the comet's orbit again next year, most meteor showers are annual events, especially if the comet has broken up enough to leave plenty of debris uniformly along its orbit. Halley's comet, for instance, produces two annual meteor showers, the Orionids in October from debris still inbound toward the sun, and the Lyrids in May from debris following its comet outbound.

But the cometary debris is often not so uniformly distributed. If it is still mostly clumped around its parent nucleus, this debris may miss the earth most years, but come back in force when the nucleus is near perihelion. The Leonid meteor shower around November 17th is usually poor, with just a few meteors per hour, but every 33 years this clumpy shower produces one of the greatest fireworks displays in nature. The 1966 display produced over 50,000 meteors per hour with the naked eye, so stress to your students to start watching this shower in November as early as 1996, and see if it builds to the expected peak on November 17, 1999. The Draconid shower, associated with Comet Giacobini-Zinner, produced fine showers in 1946 and 1992, and may do so again soon.

Let's go through the calendar for some showers to remember. The Quadrantids peak sharply on the morning of January 3rd annually and may produce over a hundred meteors per hour. The next good shower is the Perseids of mid-August, typically they give us a meteor a minute, but have been more intense the last few years as their parent comet, Swift-Tuttle, was approaching perihelion. Expect them to remain good for several years to come. The Orionids in October are the better of the two showers associated with Halley's; tell your students that if they missed the comet itself in 1986, they can see the trailers in October.

As already mentioned, the Leonid shower of mid November should soon build up to spectacular peak at the end of the decade. Closing out the best annual showers, the Geminids in mid December are remnants of Phaeton, the now dead comet masquerading as an asteroid.

All meteor showers are best seen AFTER MIDNIGHT; at sunset, the earth's counterclockwise rotation and revolution have us facing behind us, where we have been. Any meteor you see during evening hours must be overtaking the earth's 66,000 mph orbital speed. For a good analogy, consider lovebugs hitting your car while driving on the interstate. Most strike the front windshield, or the morning side of the earth. Only a very energetic and suicidal lovebug will overtake your car and smash itself into your rear windshield.

In observing a meteor shower, a dark observing site is vital; light pollution can easily wipe out all but the brightest meteors. Try to get as clear a horizon as possible; trees and buildings can easily block your view of a meteor; this is particularly important in the direction of the radiant, the constellation from which the meteor shower seems to be coming from at you. Most names are self-explanatory, but the Quadrantids in January come from the now extinct constellation Mural Quadrans, near Bootes and Arcturus.

Get warm and comfortable; a tiltable chaise lounge and a nice sleeping bag are suggested, even for the Perseids; when you are inactive, sweeping the sky with your eyes alone for an hour or more, even an August morning can get chilly; if you are too hot, you can always unzip the bag.

If you want to photograph the event, try 3-5 minute exposures with a wide angle lens (28mm or shorter f.l.) and high speed film (Konica 3200 and Fujicolor 1600 have worked well for us). As soon as you are sure a bright meteor crossed your exposed star field, stop the exposure to get as little star trailing as possible. Expect to waste a lot of film....

ANSWERS TO STUDY EXERCISES

1. A rocky body, irregular in shape, quite dark with a carbon dust coating, and pitted with many craters. The surfaces of Gaspra, Deimos, and Phobos are probably pretty typical.

2. Assuming you are visiting a comet still far from the sun which has not been close to the sun before, you would find a small body, probably like the asteroid irregular in shape and quite dark in color. The basic difference would be inside; the asteroid would be rocky throughout, while the comet would be much less dense, with lower density ices and trapped gas pockets in its interior. It is possible some asteroids also have ice, and that some comets who have rounded the sun often already have lost most of their ices and now resemble asteroids; Phaeton is of this type. Were the comet closer to the sun, the difference would be readily evident in terms of bright, active jets of gas spewing out of local weaknesses in the comet's crust. Asteroids do not show this outgassing near perihelion. Chiron shows how hard it is to distinguish. While far from the sun at discovery in 1980, it was a point of light and thus called an asteroid; as it came closer in 1992, it grew a coma and is now regarded as the largest known comet (except for Pluto?).

3. The icy materials vaporize, leaving behind the rocky stuff in the comet's orbit. This debris may strike the earth and give rise to meteor showers like the Perseids from Comet Swift-Tuttle.

4. First, no normal crustal rocks are pure iron or nickel; this combination will naturally occur only in the earth's core. If the sample is not man-made, its meteorite nature can be determined by etching it with acid. The large crystal pattern (Widmanstatten figures) found in irons is typical of VERY slow cooling deep in the core of a large, differentiated body; no man-made piece of nickel-iron can be produced with this distinctive pattern.

5. The chief problem with the nebular hypothesis is the distribution of angular momentum in the solar system. Simple nebular models predict that most of the angular momentum should still be in the sun, with little imparted out to the planets. But Jupiter alone possesses more angular momentum than the slowly rotating sun. There must have been a mechanism to transfer momentum outward.

6. Pluto certainly breaks the rules in terms of its orbit, moving at a seventeen degree tilt to the ecliptic and with an eccentricity of about .25, so elliptical it cuts inside of (but above) the obit of Neptune from 1979 through 1999; this orbit is more like that of a comet than any other planet's orbit. Also, Pluto certainly is neither a dense, rocky terrestrial, nor a low density gas giant far from the sun. With its mixture of both rock and ice, Pluto is probably most similar to a smaller version of Jupiter's moon Callisto, similar to it in density.

7. Jupiter is farther from the sun than the earth, and thus formed in a much colder part of the solar nebula. The major difference between them now is the high portion of volatile hydrogen and helium Jupiter retained, but the earth and other terrestrials lost due to the sun's heating.

8. The rotation of the solar nebula flattened the solar nebula before the planets condensed out; by the time they accreted, the dusty disk was flatter than a pancake, and the planet orbits (except for Pluto's, anyway) reflect this by staying in the ecliptic; this also closely matches the sun's equatorial plane.

9. Giotto found Halley's nucleus both very black and dirty (carbon grit) and very icy, as shown by the active jets of gas where the ice was exposed to the sun's rays. Rather than the "snowball", however, the nucleus was quite elongated and not round at all.

10. First, how dense is it? Irons are about three times denser than common crustal rocks. When cut and etched, irons also will show large crystal patterns not found in any specimens from here.
If the specimen is filed, and shiny iron flecks appear, it may be a stony iron or even some type of stone. These are not nearly as dense as irons, however, and are easily confused with many crustal rocks also containing iron. If the fall is quite fresh, you may still see a dark fusion crust, melted by the heat of entering our atmosphere. But some oxidation and weathering processes on earth can give rocks similar appearances, so it is always best to check with a local university; astronomers and geologists there may be able to confirm its meteorite status, or return if to you with the notation that it is a "meteorwrong;" if they are unsure, they will send it on to the American Meteorite Society for still more analysis. Our record at Pensacola Junior College is mixed. We helped confirm one spectacular meteorite; the Grayton Beach Meteorite, discovered in the fall of 1983, is the largest specimen found in the Southeastern U.S. since 1957, as confirmed by the American Meteorite Society. We have also had six "wrongs" to date--but a lot of fun is in the hunt!

11. As the bodies in the asteroid belt collided, much of their debris probably fell in toward the sun and hit the terrestrial planets.

12. Three processes aided in this heating. Accretion implied collision with smaller debris; the larger the body became, the greater its gravity, and thus more violently the smaller bodies impacted it and thus melted portions of its crust. As it grew, gravitational contraction shrank and heated the body's interior, melting it and causing differentiation. Also, the denser core became enriched in heavy elements, some of which, like Uranium, are radioactive. Their decay caused the continuing heat flow upward that causes volcanoes, quakes, and plate tectonics.

13. Gigantic Jupiter (and probably Saturn as well) may have begun as protostars, already much hotter than the smaller planets. They were perhaps even "brown dwarfs" for a while, giving off light during their rapid early gravitational contraction. The increasing density of Jupiter's moons as you approach Jupiter suggests it caused them to mimic the density of the planets around the sun.

14. The heating of the nebula is due to gravitational contraction and collisions between the particles circulating around in the disk.

15. Due to angular momentum, the spinning will flatten the disk into the ecliptic plane.

ANSWERS TO PROBLEMS & ACTIVITIES

1. To estimate the mass, we need to know its density and volume. Ceres has a radius of about 500 kilometers. If it is all rock, its density might be close to 3000 kg/cubic meter. then its mass would be mass = density x volume = 3000 kg/m^3 x 1.33 x (3.1416) x (5.0 x 10^5 m)^3 = 1.6 x 10^{21} kilograms. Its escape velocity is (2GM/R)$^{.5}$ = (2 x 6.67 x 10^{-11} x 1.6 x 10^{21} / 5 x 10^5 m)$^{.5}$ = 650 meters/second, less than a tenth the earth's.

2. As icy materials have a density of about 1000 kg/m^3, we solve in the same manner as in problem 1 above, and get its mass = 1000 x 4.2 x (10,000 m)^3 = 4.2 x 10^{15} kg. Then calculating its escape velocity, we get (5.6 x 10^5/10^4)$^{.5}$ = 24 m/sec, far less than that for Ceres.

3. Solving for the combined masses, M = 4 x (3.14)^2 x (100 m)^3 / 6.67 x10^{-11} x (14,400 sec)^2 = 2.86 billion kilograms.

4. If its aphelion is 100,000 A.U. and perihelion lay close to the sun, then its average distance from the sun is about 50,000 A.U. Using Kepler's Third Law, we get P^2 = a^3 = (5 x 10^4 A.U.)^3 = 1.25 x 10^{14}; thus P = 11.2 million years.

5. To resolve details down to 300 meters at a distance of 2.5 million km would require resolution of .025 arc second!

6. Ceres orbits about 2.8 A.U. from the sun, or at opposition it is about 1.8 A.U. from earth. It its diameter is 960 kilometers, then its size is sin x = 960 km / 1.8 x 1.5 x 10^8 km = 3.5 x 10^{-6}, so the angular diameter is .73 arc seconds, still a good bit larger than Pluto's disk.

CHAPTER 13

Our Sun: Local Star

CHAPTER OUTLINE

Central Question: How does the sun produce its life-giving energy, and how does its energy production and flow affect its physical characteristics?

13.7 The Active Sun
   A. Sunspots
   B. Sunspot cycle
   C. Physical nature of sunspots
   D. Prominences
   E. Flares
   F. Coronal loops and holes
   G. Model of the solar activity cycle

Enrichment Focus 13.1: The Sun's Luminosity

Enrichment Focus 13.2: Radiation from Blackbodies

CHAPTER OVERVIEW

Teacher's Notes

While aspects of the quiet sun will apply to studies of other stars, it is the active sun that we see when observing it in detail (both in space and time). And it is often these details that students are somewhat familiar with (sunspots and CB radios, for example). A survey of all solar features, backed up with photographs, will illustrate what we do observe on the sun, and just how complex the problem of understanding the sun is in detail. Many slides and films of solar features are available. When using them, it is often more meaningful to relate sizes of the features to the earth's size and to the earth-moon distance. For example, a large sunspot is about the same size as the earth. The earth-moon distance would stretch about a third of the sun's diameter across the disk. If lucky, you might even find a large spot, trailed by a much smaller one about a third of the disk away, to silhouette the earth-moon system.

Emphasize the importance of magnetic fields for all aspects of the active sun. Note that the sun must have an internal dynamo to generate its magnetic field. BE SURE THE STUDENTS REVIEW AND UNDERSTAND THE MATERIAL IN CHAPTER 5 BEFORE THEY PLUNGE THROUGH THIS CHAPTER!

Most students have heard the terminology, but do not understand the characteristics and processes of solar features. For example, prominences are often confused with solar flares. Such these terms appear in the news from time to time, it would help to define and classify features accurately. A lab experiment on measuring the solar luminosity may be done easily. Place a blackened tin can filled with a known amount of water in sunlight. Measure the temperature at regular intervals over about two hours. Plotting temperature as a function of time give s the rate of energy absorption by the water. Energy loss form the container and water can also be determined experimentally. The net rate of energy gain is the solar constant. Include in the discussion why the value of the luminosity may be considered a LOWER limit to the luminosity, and try to make guesses or approximations to correct the experimental value. It would be advisable to do a sample calculation relating the detector area to the area of a sphere the size of the earth's orbit. Don't expect non-technical students to know formulas for the area of a sphere or to be able to handle ratios.

If possible, observe sunspots. Caution students about the hazards of observing the sun visually, and point out that projecting the image of the sun onto a screen is the safest method, as well as most practical for a whole class. We had several students extend sunspot observations over several days, on their own, to determine the solar rotation rate.

Emphasize the role that models play in understanding the solar interior. Here is an example of a model which is thought to be fairly reliable and that has led to specific predictions. The continued failure to observe or verify the predicted solar neutrino count by observatories world-wide is forcing astrophysicists to reconsider features of the model. Discuss possible solutions with your students, such as the sun's interior being cooler than predicted, or solar neutrinos having a tiny mass and possibly decaying in transit.

13.1   *A Solar Physical Checkup*

That the sun is a star is one of the first fundamental concepts young people must grasp as they begin to understand the universe around them. We know it well, for we are its children; our eyes are tuned to its peak energy, and many societies worshiped the sun as a god. It is close enough that any optical aid will reveal detail on its fiery surface. Yet still our knowledge of our home star is far too inadequate to suit us.

For a quick look at the sun's volume, consider that the sun is almost 100x larger in diameter than our planet. Hence in terms of volume, the sun is $(100)^3$ or a million times larger; you could pack a million earth's inside the sun globe. But the sun does not weigh a million times more than earth; its mass is only about 333,000x greater; the sun is 1/3rd our density.

13.2   *Ordinary Gases*

If your students are unfamiliar with Kelvin, note that room temperature is about 295 K, and that each Kelvin degree is equal to 1.8 Fahrenheit degrees. In the unit on stars, with temperatures of thousands of degrees, the 273 degree difference between the Celsius and Kelvin scale is hardly noticeable; approximate Fahrenheit degrees by doubling the Kelvin temperatures.

13.3   *The Sun's Continuous Spectrum*

In considering the sun as a blackbody, remember the term refers to color black's ability to absorb all the other wavelengths falling upon it. But that energy is conserved; the body itself grows hotter from the absorbed waves, and now reradiates this energy as a continuum peaked in the infrared (i.e. as heat waves). For graphic proof, who is warmer on a cold but clear winter day, someone wearing black or white clothing?

In discussing the sun's opacity, remember that in the sun' denser interior, energy is carried upward via convective currents and atoms colliding with thermal radiation. The photosphere is that point at which the density drops low enough for the energy to escape as radiation. The photosphere's density, incidentally, is about 3,400x less than that of the air you are now breathing, yet it is still dense enough that most of the radiated energy leaves the sun in the form of the familiar continuum we call the rainbow of colors.

13.4   The Sun's Absorption-line Spectrum

But certain atoms are at the proper temperatures (particularly sodium, calcium, and iron) that they do create absorption lines in the photosphere, accounting for the dark lines we observed with the spectroscope in the sun's spectrum in chapter nine. But remember that these are just the elements with obvious absorption features about 6,000 K; helium does not show up well in the sun's spectrum, despite being the second most abundant element. Helium is a noble gas, and hangs onto its electrons tightly; as we will find in later chapters, helium shows up much better in the spectrum of hotter stars, where the higher temperatures can excite or even ionize the stubborn helium atoms.

13.5   Energy Flow in the Sun

In considering the convective flow up to the photosphere, consider that the granules are larger than the great state of Texas, yet so turbulent is the upward flow that they have dissipated in the eight minutes it took for their radiation to reach us.

In the nuclear reactions at the sun's core, gamma rays are the chief form of energy produced. They at first move freely outward in the radiative zone, but are then absorbed by the opaque gases of the convective zone. Here collisions between atoms dissipate the energy of the gamma rays with millions of subdivisions. By the time a gamma ray's energy reaches the photosphere, it may have given birth to six million photons of visible light! That's good for us, for we had much rather receive light than gamma rays from our star.

If the solar corona is so much hotter than the photosphere, some students may wonder why we must wait for a total solar eclipse to observe it. The answer is two-fold. First the high temperature means that the radiation peak is shifted far shorter than visible waves; the corona in fact is the sun's primary source of X-rays. But as the corona is typically 500,000x thinner than the photosphere, this means that there are few atoms about there to radiate that energy. By shear weight of numbers, the 6,000 K atoms of photosphere dominate the sun's radiated energy.

13.6   The Solar Interior

At the turn of this century, only two ways of making energy on the scale the sun is doing were known, and both seemed totally inadequate to explain the longevity of the sun and the life it has fostered on earth. Gravitational collapse from a nebula the size of our orbit would produce solar energy for only about 35 million years, and the chemical combustion of a sun-sized lump of carbon would take even less time. The discovery of the strong nuclear force allowed us to visualize a force far more efficient in converting matter into energy than any chemical reaction, as noted in the megaton discussion in chapter nine. In the 1920's Eddington suggested that the fusion of hydrogen to helium provided solar energy; few great ideas have been accepted so readily, for everyone knew that gravity and chemical reactions could not explain the great energies generated by the sun and other stars.

In conjunction with Einstein and matter into energy conversion, James Bond fans will be delighted to note that .007 of the mass of the four hydrogens ends up as gamma ray energy, rather than as helium; this is true in both the proton-proton and CNO cycles. Both processes produce exactly the same products in matter and energy; in the CNO, the carbon's presence acts as a catalyst to speed up the reaction.

What happens to the positron? Star Trek fans know that antimatter is the fuel for the Enterprise, and it is annihilated when combined with normal matter. When the position meets an electron (doesn't take long, there are plenty of free electrons in the hot plasma in the sun's core, both particles are obliterated, and two gamma rays fly off in place of them.

Note that the sun is only about half way through its hydrogen burning stage. No need to run out and buy red giant insurance yet...

While the solar neutrino problem continues to baffle us, my own bet is that the neutrino will turn out to be unstable. Perhaps in the eight minutes from sun to us, over half of the neutrinos produced in the sun's core decay into some presently unobserved form. Stay tuned.

13.7   The Active Sun

The sunspot cycle is not exactly eleven years. It has been as short as eight years and as long as fourteen years. Our current cycle seems to be on the short side; the peak was probably in early 1991, or about a year earlier than predicted, based on the last minimum.

While the sunspots look dark against the bright background of the yellow hot photosphere, 4000 K is by no means cool. If plucked off the sun's surface and hung in space, the individual spot would be as bright orange as Arcturus of the spring sky, and brighter than the full moon!

People have often tried to tie earthly events to the eleven year sunspot cycle with little success. But as the Planet Earth video shows, we are getting much better results with the 22 year polarity reversal cycle and terrestrial climatology, particularly with El Nino and droughts.

Zeilik does not mention the Maunder Minimum, but it certainly is worth noting. For seventy years, from 1645 until 1715, observers with scopes far better than Galileo's observed the sun and found few if any spots. As sunspots are cooler regions, you might at first think their absence would make the sun and earth warmer. But climatic records paint a different picture; in Europe this was the "little ice age", when the Baltic froze over for months every winter and a regular horse-drawn sleigh service carried passengers and freight across it from Stockholm to Copenhagen weekly every winter. Sunspots' cooler temperatures are more than made up for by the extra activity surrounding them; the sun is at its warmest near sunspot maximum.

As sunspots are certainly magnetic phenomena, I like to compare them to magnetic shields. A strong field will deflect the normal upward convective flow of heat in the granules, so the poles of the field grow colder and darker. But if the field is suddenly disrupted, and the magnetic bottle breaks, the hot gases that gush upward may be 10x hotter than the normal temperature of the photosphere. They erupt at the surface as a short-lived flare, spilling out all forms of radiation which in a matter of minutes can disrupt communications and threaten the lives of astronauts in deep space (see discussion below).

ADDITIONAL RESOURCES

Transparencies 13.2, 13.4, 13.5, 13.6, 13.12, 13.16, 13.17, 13.21, 13.23, and 13.31 will assist you with this chapter.

The "Solar Ocean" episode of the Planet Earth video series is most highly recommended. It ties the sun's energy and variability in well with life and climatic changes on the earth and is a clarion call for more research on our vital home star.

SUPPLEMENTARY ARTICLES

1. "The Rotation of the Sun," <u>Astronomy</u>, December 1986, page 106.
2. "Solar Prominences," <u>Astronomy</u>, July 1987, page 18.
3. "Solar Flares," <u>Astronomy</u>, November 1987, page 18.
4. "The Inconstant Sun," <u>Astronomy</u>, October 1988, page 22.
5. "Safe Solar Observing," <u>Astronomy</u>, February 1989, page 66.
6. "The Violent Sun," <u>Astronomy</u>, February 1990, page 32.
7. "Where are the Solar Neutrinos?," <u>Astronomy</u>, March 1990, page 40.
8. "The Fiery Fate of the Solar System," <u>Astronomy</u>, April 1990, page 20.
9. "An Observer's Guide to Sunspots," <u>Astronomy</u>, May 1991, page 62.
10. "One Day on the Sun," <u>Astronomy</u>, January 1992, page 48.
11. "Observing Solar Flares," <u>Astronomy</u>, February 1992, page 74.
12. "Fire in the Sky--Observing Auroras," <u>Astronomy</u>, March 1992, page 38.
13. "Heating the Solar Corona," <u>Astronomy</u>, May 1992, page 26.
14. "Journey to the Heart of the Sun," <u>Astronomy</u>, January 1995, page 38.
15. "Discover our Daytime Star," <u>Astronomy</u>, February 1995, page 66.
16. "Bands, Glows, and Curtains," <u>Astronomy</u>, April 1995, page 76.

ADDITIONAL DISCUSSION TOPICS

Why don't we speak Norwegian?

Had you been a betting man in 1000 AD, you no doubt would have favored the Norse to conquer and colonize the New World. They built the best ships afloat, and their brave navigators had already sailed to and colonized Iceland and Greenland. And now came news of Vinland's discovery. Already colonists were signing up to go west yet again. And under the leadership of Leif Erickson, they did return to Vinland time and again.

The Viking trading posts in the New World were intended to follow a pattern that civilized the largest nation on earth back in the Old World. The Russ were Viking traders, establishing trading posts along the river network of central Europe and Asia. When they ran out of rivers, they portaged their long boats to the next river and headed back downstream. In this fashion they established Novograd, Moscovy, and Kiev, and explored the Volga and other rivers of European Russia. Some speculate that Viking traders not only built along the coast of eastern Canada and the northeast United States, but even ascended into the Great Lakes much as they did in Russia.

But this concept needed constant resupply of trade goods, food, and weapons from Europe to succeed. The first English and Spanish attempts to colonize the New World would all fail in part for lack of adequate resupply shipments. But the Europeans after Columbus had gold fever, and they kept coming, much to the displeasure of my own native ancestors. Why didn't the Vikings show similar commitment to colonization?

The answer lay in the sun. Much like the Maunder Minimum of 1645-1715, the Sporer Minimum of the Thirteenth Century found the solar activity declining for decades, with the earth becoming colder as a result. The Viking supply route was a far northern one, from Scandinavia to Iceland to Greenland, then south to Vinland. The critical Greenland connection was broken when pack ice blocked even the sturdy long boats from reaching Greenland during this period.

Have you ever thought about the name "Greenland"? Sounds like a real estate agent came up with it, doesn't it? One did; his name was Eric the Red, and he promised his neighbors to lead them to a land of green fields for growing sheep for wool and mutton. He was not lying; in 1000 AD the southern coast of Greenland was comparable to Ireland today, and for two centuries Greenland wool grew famous for its quality, until it became the fabric of choice for European royalty. The Pope recognized the enterprise of the Greenlanders, and appointed a bishop for the small cathedral the Viking settlers built in their capitol. The Spanish were not going to get similar papal recognition for their efforts in Hispaniola for another 400 years.

But as the weather chilled, the grass died, and so did the sheep. No more boats from Iceland appeared for years, and the desperate Greenlanders "went native", intermarrying with the Inuit and adopting their life style, better adapted to the cooling climate. Much the same course occurred in the new world as the Viking traders lost contact with their European culture.

The sun was to frustrate Scandinavian colonization a second time. This time it was the Maunder Minimum that was to stop Swedish attempts at colonizing what is now Delaware. When the Swedish ships were stuck in Baltic ports for months every winter, the desperate colonists talked the Dutch in New Amsterdam into annexing them.

The Maunder and Sporer Minimums show that the sun has poorly understood activity cycles which must take hundreds of years. These have great implications to our understanding of both history and climate. Perhaps Joseph's seven lean years have an astronomical basis. Maybe the downfall of the Roman empire was more related to lack of solar activity than to the new activity of the Huns and Visgoths, after all.

Have your class consider the impact of a new Maunder Minimum on geopolitics right now. Colder climates would hurt the Russians badly, as well as the Canadians and farmers on our Great Plains. But increased rainfall might cause the deserts to bloom; certainly the Sahara was 10,000 years ago far greener than the desolate wasteland we see today. We must learn to look beyond national boundaries, to our planet as a whole, if humanity is to thrive here.

Solar Flares and Human Space Exploration

Some of your students might have read James Michener's epic, <u>Space</u>. At its conclusion, the fictional Apollo 18 mission comes to a disastrous finale when two of the three astronauts are trapped on the moon's surface by the intense radiation of a strong solar flare. This is not science fiction...it will happen someday. It was through good luck that it was not a problem in any of the real Apollo missions, but remember they only took less than two weeks. It is not now a problem for space shuttle astronauts and Mir cosmonauts because all of these are in low earth orbit, well below the earth's protective van Allen radiation belts. But if we go up even to geosynchronous orbit, we leave the earth's magnetic force field behind us. Certainly any colony on the moon will need a "fallout shelter" where the colonists can retreat to ride out the worst radiation from a major solar flare. Likewise, a Mars-bound expedition will need ships equipped with a lead-lined room, capable of shielding the humans from the energetic protons and electrons of the solar wind during the most intense solar activity. On Earth, we enjoy these as the best auroral displays.

For more on current research in solar problems and activity cycles, see Ken Langs' articles, "Unsolved Mysteries of the Sun", in August (part I, page 38) and September (part II, page 24) issues of <u>Sky and Telescope</u> in 1996.

DEMONSTRATIONS AND OBSERVATIONS

Observing the Sun Safely

The tips mentioned back in chapter one on observing solar eclipses apply to watching the sun anytime. NEVER view the sun without adequate safety filters. For in class viewing, eyepiece projection is hard to beat; project the sun's image through an unfiltered telescope onto a piece of stiff white paper, and focus the eyepiece until the edge of the sun is sharp and detail such as sunspots are noted on the disk. If you wanted to do this over several days, the students could note from sketches of the sunspot positions just how the sun was rotating, and find its period of rotation was about a month; individual sunspots remain on the hemisphere of the sun facing us for about two weeks at a time.

If finer detail is desired, you need to buy an aluminized mylar or iconel filter to fit over the FRONT of the telescope tube--filters that screw onto the eyepiece are NOT recommended. This prevents heat from ever building up in the scope and damaging the optics, much less your eyesight. Check out the latest issues of <u>Sky & Telescope</u> or <u>Astronomy</u> for dealers; Thousand Oaks makes good quality filters of both types, as does Roger Tuthill. Since the daytime atmosphere is quite turbulent due to solar heating, I do not recommend buying larger than a 4" filter; for bigger scopes, buy a mask and mount the filter off-axis and you will still get plenty of light and all the resolution your seeing conditions will permit. This will also save your school a lot of money and will not be as heavy a weight on the front of your scope to have to balance.

If you want to make a photographic record of the sunspots, get a camera adapter that will allow you to replace the eyepiece with a 35mm SLR body. Refocus the scope until the sun is focused on the viewing screen of the camera, and use either autoexposure, or a variety of fairly short time exposures (with ASA 100 speed film, about 1/500th or 1/250 second) to capture the entire solar disk. If well focused and properly exposed, you will get a very clear picture of the sunspot activity day by day.

Observing Solar Activity with Hydrogen Alpha Filters

While the visible light sun usually appears fairly quiescent (I did observe a moon-sized sunspot entirely vanish in under an hour once), the sun viewed through hydrogen alpha is far more interesting, as figure 10.15 shows. These narrow band pass filters transmit only the light of the hydrogen alpha line, and the narrower the band pass, the finer the detail you can observe on the disk (but the more expensive the filter will be to make and purchase). They can run thousands of dollars, but I think a good value is Lumicon's Solar Prominence Filter for around $600; check out the current issues of <u>Astronomy</u> and <u>Sky & Telescope</u> for price updates.

The filter has two parts; an energy rejecting prefilter that blocks all but deep red waves from entering the telescope, and a tunable narrow band pass final filter placed in front of the eyepiece. By altering the tilt of the final filter, you can optimize the contrast and view detail along the sun's limb in amazing detail. Note that these filters are sensitive to changes in temperature; you must occasionally retune them as the filter warms up or cools off. So if you are showing your entire class an eruptive prominence, and a student claims he can't see it anymore, quickly take a peak to be sure the scope is tracking correctly, or that the filter doesn't need retuning, or (and this can happen in a few minutes) that the prominence hasn't actually faded that fast.

For that reason, it is a good idea to keep a camera handy to attach after the final filter and photograph some prominence activity as it is happening. A sequence of shots over an hour will show quite a bit of change. In fact, if you try to sketch the prominences, you must work VERY FAST; they can literally change shape right before your eyes.

Be warned that watching the sun through hydrogen alpha filters is addicting; once my class began the period by looking at the sun in hydrogen alpha, then we saw a "white light" flare (visible even on the normal projected image of the sun) erupt, with a prominence shooting upward from it. We blew the entire period watching the changes in the prominence as it arched into space, then followed the magnetic field lines back downward in a graceful loop. For my students, this experience was anything but a wasted class, however. The testimony to the power and variability of our star left them in awe.

To observe still finer detail in the chromosphere (remember it is the red edge of the sun as totality begins and ends), you need a still more narrow band pass, something in the range of .8 to .6 angstroms. These filters must be stabilized with an oven for precise temperature control, but do reveal the magnetic field lines around sunspots and a host of other detail on the disk that the Lumicon filter does not show. They also show solar flares, especially the frequent smaller ones, far more vividly than just the prominence filter can. But they typically cost three to five times more than the solar prominence filter mentioned. You know your own school's budget--get the best you can afford.

Observing Differential Rotation

You can spot differential rotation if you have two sunspot groups on the sun at different latitudes in the same hemisphere. Continuing observations will show spots near the equator rotating in about 25 days, while ones near 30 degrees latitude above or below the equator will take about five extra days to rotate. As most sunspot groups do not last long enough for you to observe this, it could take some luck and about a month's worth of observations to show this.

ANSWERS TO STUDY EXERCISES

1. The planet Venus reflects radar well, but the sun does not reflect radio waves well. We can determine the distance of Venus from the sun very accurately via Kepler's laws, and the radar signal gives us a precise measurement of the distance between us and Venus.

2. We need to know the earth-sun distance in units such as meters (see question above). Then we also need to know how much energy from the sun reaches the earth very second over some surface area (such as a square meter). This measurement is best made from space, because the earth's atmosphere absorbs some of the incoming sunlight, giving a lower value than the actual one. The earth's atmosphere absorbs some of the solar radiation at wavelengths shorter and longer than the visible region of the spectrum. The earth-sun distance can be measured with great accuracy, but the sun's flux is not as well known.

3. The solar spectrum appears as a continuous band of color from red to violet, with many dark lines appearing against this colored background. Some of the lines appear wider or darker than others. The darkest lines are those of ionized calcium, atomic sodium, and iron. Astronomers call this an <u>absorption</u> spectrum.

4. An absorption spectrum occurs when electrons, in discrete energy levels, absorb photons. That requires that the electrons rise up in energy level. When they drop back down, they emit photons in any direction, compared to the photons absorbed along the line of sight. Also, the photons emitted as the electrons go back down may be at different energies. For instance, the blue-green Balmer line of hydrogen absorption is created by the electron jumping up from level 2 to 4, but the emission lines may be given off in the 4 to 3 and 3 to 2 transitions, the first of which is invisible (in the infrared) and the 3 to 2 drop is the red hydrogen alpha line. Thus the number of photons reaching the observer on earth is decreased at the wavelength of the observed absorption lines, creating the dark lines from energy absent from the continuum.

5. Helium is an inert gas, tightly holding onto its electrons. Thus it takes more energy and higher temperatures than we find in the sun's photosphere to excite helium's electrons and create spectral lines for helium. However, the hotter chromosphere can create helium lines, and this is where helium (element of helios, the sun) was found during solar eclipses during the nineteenth century, some thirty years before it was found here on earth.

6. Nuclear fusion in the core is converting four hydrogen atoms into a single helium. Thus the amount of hydrogen in the core is decreasing while helium is increasing.

7. The sun's continuous spectrum is similar to a blackbody curve, where the shorter the wavelength of the peak of the curve, the hotter the temperature of the blackbody. The sun's peak in the yellow would correspond to a blackbody temperature of about 5,800 K.

8. It takes photons produced in the sun's core, traveling at the speed of light, tens of thousands of years to reach the photosphere. This is because the interior is opaque. That means that a photon travels only a short distance before it is absorbed and reemitted--most likely in a different direction. So the photon follows a "Drunkard's walk" out of the sun, in contrast, neutrinos, traveling at the same speed, reach the solar surface in about one second. This is because that the interior is transparent, allowing them to fly out at the speed of light.

9. In its core, the sun produces its energy by fusing four hydrogen atoms into a single helium.

10. Like the rest of the photosphere, sunspots radiate like blackbodies. Isolating just the light from a sunspot, we find its curve peaks in the orange, at longer wavelengths and lower temperatures (about 4,000 K), according to Wien's Law.

11. Helium is hard to ionize; the chromosphere is hot enough to excite and ionize it, but the cooler photosphere is not hot enough to knock electrons from this very stable element.

12. We see photospheric granulation, with marks the top of the convective cells that are carrying energy upward from the solar interior.

13. Because neutrinos can travel freely through the solar interior, they tell us about the rate at which the core is converting hydrogen into helium. However, to date, the number of neutrinos actually observed is only about a third as many as predicted.

14. The pressure in the sun's interior is much greater than at the surface. This high pressure is critical to forcing the hydrogen atoms together energetically enough to initiate fusion.

15. The pressure will increase by 2 x 3 = 6 times greater.

ANSWERS TO PROBLEMS & ACTIVITIES

1. The sun's luminosity is 3.86×10^{26} Watts from fusion reactions, or 3.86×10^{26} J/sec. So every second's worth of energy comes from $E = mc^2$; the conversion into mass = $3.86 \times 10^{26} / (3 \times 10^8)^2$ or the sun converts 4.29×10^9 kilograms per second.

2. Five billion years = 1.578×10^{17} seconds, so in this time, the sun has converted 6.78×10^{2} kilograms into radiated energy. This only represents .00034 of its entire mass, however.

3. Wien's Law gives the peak wavelength = $2.9 \times 10^{-3} / 300$ K = 10^{-5} meters = 10 micrometers. The flux emitted by every square meter of the earth's surface is $5.67 \times 10^{-8} \times (300)^4 = 460$ W/m^2

4. Since your body temperature is close to 300 K also, you are emitting energy at about the same rate as the earth in # 3 above. The sun is emitting about 6.3×10^7 W/m^2 (see focus 13.1), so the sun is giving off about 137,000 X more energy than your body does. As the sun is about 19.3X hotter than your body, Stefan's law gives $(19.3)^4 = 139,000$ X, in good agreement with this calculation.

5. As the sunspots have a temperature of 4200 K, Wien's Law gives their peak wavelength as $2.9 \times 10^{-3} / 4200$ K = 6.9×10^{-7} meters, or 6900 Angstroms, at the very reddest end of the visible spectrum.

6. At 1 AU, the sun's luminosity has spread out over an area of $4 \times 3.14 \times (1.5 \times 10^{11}$ m$)^2 = 2.83 \times 10^{23}$ square meters. The earth's surface area is $3.14 \times (6.378 \times 10^6$ m$)^2 = 1.28 \times 10^{14}$ square meters, or we intercept about 4.5×10^{-10} of the sun's total radiated energy.

CHAPTER 14

The Stars as Suns

CHAPTER OUTLINE

Central Question: How do astronomers determine the physical properties of stars?

14.1   Some Messages of Starlight
      A. Brightness and flux
      B. Flux and luminosity
      C. The inverse-square law of light

14.2   Stellar Distances: Parallaxes

14.3   Stellar Colors, Temperatures, and Sizes
      A. Color and temperature
      B. Temperature and radius
      C. Direct measurement of diameters

14.4   Spectral Classification of Stars
      A. Temperature and the balmer lines
      B. Spectral classification

14.5   The Hertzsprung-Russell Diagram
      A. Temperature versus luminosity
      B. Spectroscopic distances

14.6   Weighing and Sizing Stars: Binary Systems
      A. Types of binary stars
      B. The doppler shift for binary stars
      C. Eclipsing binary stars
      D. Sirius: a binary system
      E. The mass-luminosity relation for main-sequence stars
      F. Stellar densities
      G. Stellar lifetimes

14.7   Stellar Magnetic Activity

Enrichment Focus 14.1: Flux and Magnitude

Enrichment Focus 14.2: Heliocentric (Trigonometric) Stellar Parallax

CHAPTER OVERVIEW

Teacher's Notes

This chapter shifts from the local star to consider the physical nature of other stars, with the sun as a basic model. This solar-connection is essential to our understanding of the cosmos beyond the solar system. One of the more recently realized inferences from this is that stars have activity cycles.

A main idea in this chapter is distance-measuring techniques; so often students will be awed by the immense distances astronomers work with but will not have considered how those distances are measured. You might begin discussion by posing the problem of measuring the distance to any object that you can't reach. Emphasize the conceptual basis, rather than mathematics. Parallax is perhaps the easiest to explain; point out that we judge distances in everyday activities by parallax (binocular vision). Note that the students will be tempted to confuse the inverse relation for parallax (the larger the parallax, the closer the object) with the inverse-square law for light intensity.

The whole idea of magnitudes is confusing, and unfortunately the mathematics is beyond most introductory students, regardless of which approach you might take. A historical approach, or one that matches the magnitude scale to the response of the eye, would be most easily visualized. One numerical approach that might be understood is to simply define a unit magnitude as a change in apparent brightness by a factor of 2.5 You will note that, except for the discussion in Focus 14.1, I (Michael Zeilik) HAVE COMPLETE ELIMINATED MAGNITUDES FROM THE BOOK because introductory students find them so confusing. In their place, I use fluxes, which is the physical unit for the same concept.

The inverse square-square law for light is also difficult, beyond the "twice as far, one-fourth as bright" level. Beginning students often find this one of the hardest ideas to apply quantitatively, especially when comparing two light sources. Stick to the basic idea that farther means fainter. Simple examples again will be helpful. you might try demonstrating this with small light courses in a darkened room. This is a good place to remind students that Newton's Law of Gravitation is also an inverse-square law, so they have seen the same functional relationship in a different physical context.

14.1 Some Messages of Starlight

Once we realize the other stars are suns, then we next ask how they compare to our home star. To learn this, we need to know their distances, their luminosities, their temperatures, their masses, etc. We wish nothing less than a census of the stars, with all their vital statistics.

Zeilik notes that stars like to sun emit most of their energy in the visual range. This is no accident; our eyes are tuned to our star. What the sun makes best, yellow light, lies right in the middle of the visual range for humans. Had we evolved around a cooler star, we might have larger eyes, made to focus in the infrared better, since a cooler star would produce more heat than light.

But most stars are not the same temperature as the sun. Wien's Law tells us that the peak of the blackbody radiation depends inversely on the star's temperature. If we were to double the sun's temperature, instead of the peak lying in the yellow portion of the visible spectrum at about 5,000 A, the peak would shift all the way into the ultraviolet, at 2,500 A; likewise halving the sun's temperature would lengthen the peak out to 10,000 A, well into the infrared.

In fact, only in the narrow range of between about 5,000 K to 7,000 K can we agree with Zeilik's assertion that stars emit most of their energy as visible light. As we have already witnessed so often, what we see is the proverbial tip of the energy iceberg. In discussing the total radiation across the spectrum, in other texts you will run across this as the bolometric magnitude of the star.

For a feeling of the effect of the inverse square law, ask your students how light the sun would appear at Saturn's distance of 10 A.U. At 10x the earth's distance, the sun appears only $1/(10)^2$, or just 1/100th as bright as seen from earth.

14.2   Stellar Distances: Parallaxes

As Zeilik points out, we can presently get reliable parallaxes only out to about .01 arc seconds, or a limit of about 300 light years on page 306. This means when you read that a star is 500 light years distant, at present this is just a best guess. Current astronomy texts list Betelguese, for instance, as anywhere from 300 to 1,400 light years distant! Doubtless many of these distance estimates will be vastly revised in the next several years as data already being gathered by ESA's Hipparchos Astrometric Satellite Observatory and the Hubble Space Telescope are released. These get far better precision, since they are operating above the earth's murky and shifting blanket of atmosphere. Remind your students that stars do not twinkle if observed from space. The twinkling is the distortion of the star's narrow beam of light as it passes through the turbulent layers of atmosphere. Note that stars just rising twinkle much more vividly than when they climb higher later in the evening.

To visualize just how much our knowledge of the stars will be revised by these satellite observatories, with 10x better resolution, they will push the limits of good data out to stars up to 3,000 light years distant. This is 10x farther out, and will cover a volume of space 10^3, or over 1,000x as many stars as we now have good data on. Presently we do not have good distances on over half of the naked eye stars; by 1995, we should have much better data on every naked eye star.

14.3   Stellar Colors, Temperatures, and Sizes

A good example of stars of very different sizes in the winter sky is Sirius, the Dog Star, and its white dwarf companion, the Pup. The Pup is 10,000x fainter than the brighter star, but the Pup is actually a little hotter. If they were the same temperature, then Sirius has a surface area 10,000x greater than its companion, or it is 100x larger than the Pup. In fact, Sirius is just slightly larger than our sun, while its companion is about the size of planet earth. As the two are so different in brightness, and the companion lies only about 10 arc seconds from Sirius, to split them with a scope will require very good seeing (a rare commodity on winter evenings) and at least a six inch telescope and about 200x; even then, it comes and goes as the atmospheric turbulence blurs the two images. Seeing Antares in the summer is somewhat easier, but again about 200x is recommended; the color contrast between the huge red supergiant and its hotter but far smaller companion makes the smaller star appear green. Even better color contrast lies to the north, at the bottom of the Northern Cross, where beta Cygni (Albireo) is a spectacular contrast of electric blue and amber. This pair is bright enough and separated far enough to image well on a CCD and TV monitor. Remember in any case the redder star appears brighter, it also means it must be much larger than its hotter companion to equal the light output of the hotter star.

14.4 Spectral Classification of Stars

How did the OBAFGKM sequence come about? When Henry Draper began his classification sequence, he believed the intensity of the hydrogen lines was an absolute indicator of temperature, and so assigned the letter A to the stars with the darkest balmer absorption lines. But in fact the balmer lines are most obvious at 10,000 K and fade if the blackbody is either hotter or cooler. When new spectra from southern hemisphere stars revealed star bluer (and thus hotter) than any studied in Draper's original A-N sequence, they were given letter O and promoted to the head of the class. Now helium was used as the indicator of highest temperatures; it is ionized in O stars, and just excited in the B stars, which also leap-frogged past type A. Several classes now proved redundant, and were dropped, leaving us with the OBAFGKM sequence which includes about 99% of all known stars.

In helping the students remember the cooler classes, I tell my students the sun is a "good ole star" to remind them it is a G star. Within each type, we assign a number suffix to tell how far along that type it falls; the sun is G2, or 2/10ths from pure type G to type K. In a terrible pun, a star almost cool enough to be type M would not be the Dog Star (Sirius is type A1), but it is still a K9 (bad, but memorable pun). Only the cooler K stars and all of type M are cool enough to show molecular lines from TiO; in 1996 water vapor was discovered in the coolest sunspots, also about type M in temperature. Hotter stars can not form molecules, as the colliding atoms are so energetic that they would break any chemical bonds that might have formed.

14.5 The Hertzsprung-Russell Diagram

The H-R diagram started with the simple question, is there any relationship between how hot a star is, and how luminous it is? If you ask your students this, most will correctly respond that as a rule, hotter stars should be the more luminous; our naked eyes clearly tell us this, for the naked eye sky is dominated by bright blue main-sequence stars. This is exactly what the main sequence shows us, with its brightest members to the top left, where the vast majority of naked eye stars lie. But figure 11.12 shows us that while all but six naked eye stars are more luminous than the sun, among its near neighbors the sun fares much better; it is the fourth most luminous of the 40 closest stars.

Why is the main sequence "main?" The stars which fall along it include our sun, right in the middle. This vital clue tells us that like our sun, these are hydrogen burning stars, still relatively young with plenty of hydrogen still to fuse in their cores. As hydrogen is far and away the most abundant element, it makes sense that about 90 % of all stars are still hydrogen burning main sequence stars. Red giants are rare since they don't last long and represent only a small fraction of the star's main sequence life span.

Luminosity classes are not nearly as clear cut as we might hope. Since we are using just the width of the star's spectral lines, and these are blurred in part by changing atmospheric conditions, different observers may classify the same star in different classes. In the technique called spectroscopic parallax, we use the spectrum to predict the star's luminosity and absolute magnitude, then plug this estimate into the distance modulus formula (see discussion in additional topics) to work backwards to estimate the star's distance. For this reason, one current text places Betelguese 300 light years distant, another 520 light years, and yet another places it 700 light years distant; Zeilik goes still farther, making it 1,400 light years distant. Thus some would make a less luminous Betelguese only 10,000 times more luminous than our sun, yet others would make it equal to Rigel as the most luminous naked eye star, at 50,000 solar luminosities.

14.6 Weighing and Sizing Stars: Binary Systems

The playground see-saw analogy is a good one for weighing stars. The more massive companion will, like the heavier kid, sit closer to the see-saw's fulcrum (the center of mass for the binary system) and move slower as the stars orbit. As on the playground, the lighter kid sits out on the end and moves the fastest. Only the most widely separated binaries are visually resolved; their periods are typically decades or centuries. Some may take millennia, but they move too slowly for us to confirm they are actually showing orbital motion yet.

In demonstrating the spectroscopic binary system, use your index fingers on both hands to stand for a spectral line from each star. When the students see your fingers aligned, the stars are both moving across their line of sight; neither star shows us any radial velocity. A quarter of a period later, the stars appear side by side, and the lines have split most widely. One hand moves toward the class, and that line is blue-shifting. The other moves back, red shifting. Continuing the orbit, at a half period your fingers will again align. After 3/4 period, the fingers again separate, but now the hands have swapped places. After two split-recombinations, we have completed a full period. Typically, the first two stages take just a day or two and the period of the whole system less than two weeks.

In finding the sizes of stars with eclipsing binary systems, I remind students that eclipsing systems are usually close enough to also be spectroscopic binaries as well. Thus from the doppler shifts we can learn the orbital velocities of the two stars. If we found the smaller star was moving at a million kilometers per hour (not unusual for close binaries), and that it took the larger star an hour to cover it, then the smaller star is a million kilometers across, a main sequence star about the same size as the sun. If the larger star covered its companion for six hours, then the larger star is six million kilometers across, a small giant of luminosity class III.

14.7 Stellar Activity

When we note that many other stars are more active than our sun, this too has implications for life. If a variation of only 10 % in the star's temperature results in a 46 % increase in the star's luminosity ($L = T^4$, remember), then such changes could alter the earth's climate from Venus to Marslike in a matter of months; it is hard for any life form to adapt to such a wide variety of climates so rapidly. This in turn makes it easier to understand why ET has not yet come calling. Even though there are plenty of G type stars out there, ours may be among the best behaved, which is in part why we are here.

ADDITIONAL RESOURCES

Transparencies 14.3, 14.7, 14.8ab, 14.9, 14.13, 14.14, 14.15, 14.16, 14.17, 14.20, 14.21, 14.22, 14.24, and 14.27 are all essential in this chapter where diagrams are very helpful.

A good star atlas or field guide will help you pinpoint interesting stars and other celestial objects. While it was first printed almost seventy years ago, Norton's Star Atlas is still a helpful guide, with listings of brighter deep sky objects, double stars, and even notably colorful stars. Its listing of visual binaries includes notes on colors and observing tips. It is available through the publishers of Sky & Telescope.

A far more extensive constellation-by-constellation survey is Burnham's Celestial Handbook. Its listing is quite extensive, and contains brief articles on the most significant stars and deep sky objects in each region of the sky. It is available from Astronomy and Sky Publishing also.

SUPPLEMENTARY ARTICLES

1. "Epsilon Aurigae, the Largest Star?," <u>Astronomy</u>, February 1986, page 6.
2. "Star Trek: Flight to the Stars (I)," <u>Astronomy</u>, March 1987, page 94.
3. "Star Trek: Flight to the Stars (II)," <u>Astronomy</u>, April 1987, page 94.
4. "Observing Binary Stars," <u>Astronomy</u>, June 1988, page 98.
5. "Do Brown Dwarfs Really Exist," <u>Astronomy</u>, March 1989, page 18.
6. "A Star That Breaks the Rules," <u>Astronomy</u>, January 1991, page 28.
7. "Regal Rigel," <u>Astronomy</u>, February 1991, page 38.
8. "Exploring the Galactic Neighbors," <u>Astronomy</u>, October 1992, page 76.
9. "Demon Variables," <u>Astronomy</u>, October 1992, page 34.
10. "Dance of the Double Sun," <u>Astronomy</u>, July 1993, page 26.
11. "Seeing a Star's Surface," <u>Astronomy</u>, October 1993, page 34.
12. "Hypergiants: The Most Luminous and Massive Stars," <u>Astronomy</u>, March 1994, page 32.
13. "Stellar Oddballs," <u>Astronomy</u>, September 1994, page 50.
14. "Compelling Capella," <u>Astronomy</u>, February 1995, page 48.
15. "Polaris, the Code Blue Star," <u>Astronomy</u>, March 1995, page 44.
16. "Epsilon Eridani, the Once and Future Sun," <u>Astronomy</u>, December 1995, page 46.
17. "Stellar Collsions," <u>Astronomy</u>, January 1996, page 46.
18. "Cosmic Billards," <u>Astronomy</u>, July 1996, page 46.
19. "A Brown Dwarf in the Neighborhood," <u>Astronomy</u>, September 1996, page 26.

ADDITIONAL DISCUSSION TOPICS

The Sun Versus Sirius

For a graphic example of just how distant even the closest stars are, compare the two brightest stars, the sun and Sirius. The sun lies eight light minutes distant, while Sirius is eight light years distant. So the ratio of their distances is then the number of minutes in a year, or 60 x 24 x 365.25, or Sirius is about 500,000 times more distant than the sun; in most of the continental U.S, Sirius is the closest naked eye star; only in southern Texas and south Florida does alpha Centauri barely climb above the southern horizon for a few minutes on spring evenings; from Hawaii, it is considerably higher and stays up much longer.

Absolute Magnitudes

In the second century BC, the Greek astronomer Hipparchus ranked the stars by naked eye brightness into six magnitude classes, with the 20 brightest stars being first magnitude, and the faintest stars visible rated as sixth magnitude. Since then photometers have allowed us to measure the flux of stars far more precisely than Hipparchus could.

In the modern system, we define a difference of 100x in the brightness or flux of the star as a factor of exactly 100x; a star of magnitude +1.00 will be exactly 100x brighter than one of magnitude +6.00. This in turn means each single magnitude step is the fifth root of 100, or about 2.5x. Thus second magnitude Polaris is about 2.5x fainter than first magnitude Spica. Alan McRobert does a fine job introducing the magnitude system on page 42 of the January 1996 issue of <u>Sky and Telescope</u>.

The apparent magnitude is the brightness of the star as seen from earth. But the luminosity of the star compares them at a standard distance. For the absolute magnitude scale, we image all stars to be exactly 10 parsecs, or 32.6 light years away. In Hipparchus' scale, very bright objects will have negative magnitudes. The sun's apparent magnitude is -26 (as seen from eight light minutes distance), but when we push the sun out to 32.6 light years, it fades to absolute magnitude +4.78; at that distance, the sun would be just a faint naked eye star. In fact, as most naked eye stars lie more than 10 parsecs distant, and are brighter than fifth magnitude, it is obvious the sun does not compare well in luminosity to the naked eye stellar population.

The distance modulus formula compares the stars apparent and absolute magnitudes. M is the absolute magnitude, and m is the apparent magnitude. D is the distance in parsecs, and they are all related in the formula: $m - M = 5 \log D - 5$.

With this formula, if you know the star's parallax, you can calculate $D = 1/p$ and find the star's absolute magnitude. To relate this to the star's luminosity, as each step of one magnitude is a luminosity factor of 2.5x, then the luminosity L, compared to the sun's absolute magnitude of +4.78, is found by $L = 2.5^{M-4.78}$; a star of absolute magnitude +2.78 would then be 2.5x2.5, or 6.25x more luminous than the sun.

The Unseen Binaries

As Zeilik notes, the sun is unusual in being a single star; most stars appear to be members of binary or multiple star systems. But for every binary we now know of, there are probably many more yet undetected. To see two stars as separate points of light as a visual binary, they must be widely separated, implying orbital periods of decades, centuries, or even millennia. But if they move that slowly, we probably have not yet had enough observations to show any orbital motion and actually prove they are gravitationally connected at all.

To see them as spectroscopic binaries, the stars must be so close they are practically touching, with orbital periods of only a few days or weeks at most. If they are separated more widely, they move so slowly that we can not see the spectral lines splitting and shifting.

This means there is a wide range of separations and orbital periods that we have trouble studying. If we are incredibly lucky, their orbital inclinations will be exactly 90 degrees, and we can see them eclipse. But if their period is 40 years, and the eclipse lasts only days or weeks, we would have to be very lucky to see it happen at all. We have much less problem finding close eclipsing binaries like Algol that fade every few days than searching patiently for such long period eclipses. This is one area where amateurs can help; last year a German amateur caught a fairly bright star in the Orion Nebula in mid-eclipse. Millions had seen it through the telescope, but no one else had noticed it eclipsing every 40 years.

If the star is nearby, then we might, as Figure 11.19 demonstrates, detect the companion by its gravitational perturbation of the proper motion of the visible star. Several nearby dwarfs, such as Barnard's star, seem to have less massive dark companions (jovian planets, brown dwarfs?), based on such astrometric studies, although different observers have come to conflicting conclusions. Most of the possible planets announced in 1996 were detected by the masses of the planets causing periodic doppler shifts in the motions of the primary stars, but some also have astrometric data like this to support their existence. For an excellent report on this state of the art research, see the August 1996 issue of Sky and Telescope, "Other Stars, Other Planets," on page 20.

Perhaps the most exciting such unseen binaries occur when a collapsed stellar core, such as the suspected black hole, Cygnus X-1, pulls strongly on its visible companion. For such a single line spectroscopic binary, only the visible star's lines are seen, but they red shift, then blue shift as the system revolves around the common center of mass. In the case of Cygnus X-1, the cycle takes 5.6 days. From the estimated mass of 30 sun's for the visible star, this leads us to expect that the unseen companion is between 6 and 11 solar masses, well above the 3 solar mass limit for black holes.

In any case, there are a lot of undetected binaries, with periods from weeks to decades, that are just waiting to be detected as our technology improves. A lot of doctoral theses for generations to come are coming in from Hubble Space Telescope right now.

Of Big Spenders and Misers

The luminosity of the star is very dependent on the star's mass. For main sequence stars, this is approximately equal to cube of the star's mass, so a star of 4 solar masses will be about 64 times as luminous as our sun. But stars do not shine for free; such a star must be consuming its hydrogen 64x faster than the sun, yet it has only 4x more mass (fuel) than does the sun. Thus the time it spends on the main sequence will be only $(T = M/L)$ only 4/64, or 1/16th as long as the sun's ten billion years, or about 600 million years. From this, it is obvious that the bright blue stars that dominate our naked eye sky are not going to be likely abodes for intelligent life; they simply will not remain stable nearly long enough for life to progress that far.

DEMONSTRATIONS AND OBSERVATIONS

Observing Colorful Stars

Zeilik wisely starts with the color contrast of bright winter stars, but you should also mention yellow Capella (the brightest G type star in the northern hemisphere) along with orange Aldebaran in connection with the winter sky. Even if it isn't winter, there are fine examples of colorful stars out. In spring, look at the color contrast between orange Arcturus and blue-white Spica, the two brightest stars of spring. In summer, the bright summer triangle of Vega, Deneb, and Altair all appear white, but Antares to the south is almost as cool and orange-red as Arcturus. Only fall is lacking in bright, colorful stars.

Photographing Star Colors

You will be amazed by how vividly the colors of stars stand out on homemade astrophotos taken with the simplest of equipment. I recommend slide film as the most vivid means for class demonstrations, but your students might like to take prints to pass around in class. You will need a camera with a B (bulb) setting for exposure (any good SLR should have one), a reasonably fast lens (f/2.8 or lower) of medium focal length (a 50mm lens will frame Orion well, and almost get all of the Big Dipper), and a cable release to fit your camera. To hold the camera steady during the time exposure, a tripod mount is needed. All of these are items most photographers will need anyway.

Pick out a bright constellation, like Orion or Scorpius, with a lot of colorful stars. Use fairly fast film, such as Ektachrome 200 or 400, or Kodacolor 100, 200, or 400; superfast films will record fainter stars but lack the color saturation of slower films. Better yet, try Infrared Ektachrome for a view of the sky in heat, not light waves. The blue stars will fade, but cooler stars will stand out.

To capture the stars as points of light, hold your time exposure to no more than 30 seconds; the earth's rotation will cause the stars to trail for longer exposures. Be sure the lens is focused at infinity, and the shutter is wide open. A dark observing site helps, but is not vital; if the constellations are along the horizon, you might even want to include some buildings or trees for scenery.

This is just the beginning. The earth's rotation can be an asset; try using it in the technique called spot-streaking. Start with the 30 second exposure as before, but then, without touching the cable release, slip the dust cover over the lens for about a minute. Now remove the dust cover carefully, and continue the exposure for another five minutes or so. The result is a pattern of stars as points for constellation identification, then as colorful trails for temperature analysis.

Note how vividly the colors of the stars stand out on your slides or prints, compared to their lackluster appearance with your eyes. If you are shooting Orion, the pink color of M-42, the Orion Nebula, will stand out just below Orion's belt. This is that good old hydrogen alpha emission line we met in chapter nine again. In the summer, the Lagoon Nebula, M-8, can be shot just above the pour spout of the teapot of Sagittarius; it too is a bright pink H-II region.

You might next try wide-angle shots to capture more of the sky. With my 28mm lens I can capture all of Orion and Canis major and Canis minor. Such a lens in a dark site can also show the star trails vividly in exposures of several hours. Directly toward Polaris, the stars all make concentric arcs around the celestial pole. Eastward, they climb in parallel lines upward toward zenith (Orion looks really good this way). To the south, they also make arcs, centered around the south celestial pole far below our horizon in the U.S. To avoid sky fogging, keep light pollution for such long exposures to a minimum by getting away from city lights as much as possible.

For telephoto lenses and longer exposures, you will need to track the stars. An equatorially mounted, clock-driven telescope, properly aligned on the celestial pole, is an ideal platform for "piggy-backing" your camera. You can probably buy or build a simple mount for sturdily attaching your camera to your telescope. Carefully track the guide star through the telescope eyepiece during your time exposure; any interruption will result in a multitude of "double stars" or elongated images. The longer the focal length of the telephoto, the more critical the precise tracking becomes. But these longer focal length lenses can also capture many deep sky objects, such as bright galaxies, nebulae, and clusters, quite well.

Suggested Visual Binaries

While the listings in Norton's and Burnham's guides are much more extensive, there are several telescopic binaries bright enough to be easily shown to your classes, even under light polluted conditions. In the summer sky, beta Cygni is a widely separated and very colorful (orange and blue) sight at 20x. Epsilon Lyrae is the "double double", with widely separated companions easily split with binoculars at 7x, then each in turn resolved with 3" scopes at 120x; note the pairs are almost perpendicular to each other in their present alignments. To the south, Zeilik has already mentioned Antares at the heart of Scorpius; use at least a 4" scope and 200x. For autumn, my favorite is gamma Andromedae, well resolved in a 60mm refractor at 60x; the colors are subtle, and stand out better with larger scopes. In winter, Castor of Gemini fame is a bright but close system; the two equally bright white stars resolve well with a 3" at 100x. As already mentioned, Sirius is far more challenging. I have seen it with an 8" at 200x; the atmosphere must be very steady, an unusual event on winter evenings.

But even a 60mm refractor at 60x will resolve the core of the Orion Nebula, M-42, into a spectacular Trapezidium, a diamond of four young stars that lights up the nebula. Among the hottest known O type stars, this quartet's ultraviolet radiation causes the nebula to glow visually.

For spring, gamma Leonis is a nice pair for smaller scopes, but my favorite is gamma Virginis, where the two white stars are perfectly matched like car headlights at 100x in a 3" telescope. Farther south, alpha Centauri is a spectacular visual binary, easily seen at 20x with a small scope; again, only Hawaii is likely to have it high enough for easy resolution.

Watching the Demon Wink

Algol, the brightest eclipsing binary, is so well known that every month <u>Sky and Telescope</u> publishes a listing of the times of primary minima for that month; these are given in universal time (Greenwich Mean Time), so for EST, subtract five hours, and for PST, eight hours to find the local times; also remember than an eclipse on November 22 at 2 hrs UT will be seen at 10 AM on the PREVIOUS evening in CST. While often these happen in daytime, or while Algol is below our local horizon, on autumn evenings you can be pretty sure several eclipses can be at least partially observed by your students, for Algol is second magnitude at maximum light, almost the same brightness as Polaris. Compare Algol with the brightest star in Perseus, Marfak. If both appear almost equal in brightness, then the two members of the Algol system are not eclipsing and both contribute their light to its apparent brightness. But for six out of every seventy hours, the larger giant orange star covers most of its brighter blue companion, and the system appears to fade to 1/3 its normal brightness (no brighter than third magnitude kappa Persei, the faint star just SE of Algol with the naked eye); this change is easily noted with the naked eye. Photos will show not just the star fading hour by hour, but also the change in color as less and less of the brighter blue star is seen.

ANSWERS TO STUDY EXERCISES
1. Betelguese is the coolest, with Capella intermediate, and Sirius the hottest. Sirius is probably about 10,000 K, or about three times hotter than Betelguese.

2. The MI star is both largest and most luminous; it is a red supergiant like Betelguese.

3. The size of the earth's orbit (diameter of two A.U.) and the resolving power of the telescopes used to make the measurements. Presently we are limited to measuring down to about .01", or to accurate distances out to about 300 light years. If Hubble eventually reaches its predicted resolution limit of .001", then we can get accurate distances out to about 3,000 light years.

4. (a) As Capella is more luminous than the sun, but about the same temperature, it must be larger than the sun to explain its extra luminosity. (b) If Regulus and Capella are equally luminous, but Regulus is hotter, then Capella must be larger to compensate for this difference. (c) Vega is more luminous; it appears fainter because it is considerably more distant than Sirus. (d) Pollux is cooler than Vega, so would appear slightly redder.

5. As our sun's x-rays come chiefly from its corona, and some stars are considerably better x-ray sources than is the sun, this implies they have even more active coronas than our star does.

6. From a binary system, we use Kepler's and Newton's laws to find the total mass of the system, then from their individual motions divide the mass between the two stars. If it is also an eclipsing binary system, we can use their orbital velocities and the duration of the eclipses to measure their diameters. Now the density can be found by dividing the mass by its volume.

7. The spectral lines of only the brighter star would probably be visible, yet the companion's presence would still be revealed because of periodic doppler shifts in the lines of the visible star, due to its orbital motion toward and away from us with its unseen companion.

8. A star of five solar masses will be about 250X as luminous as the sun.

9. By the inverse square law, the flux at Jupiter is $1/5^2$, or just 1/25th as great as at the earth.

10. If the color is identical, so are their surface temperatures.

11. Supergiants lie about the top center to upper right of the H-R diagram.

ANSWERS TO PROBLEMS & ACTIVITIES
1. As 1 pc.=206,265 A.U., then the sun appears $1/(206,265)^2 = 2.35 \times 10^{-11}$ times fainter.

2. If the star were five times more distant, its flux would drop to 1/25th the original.

3. As Mars has an orbit 1.52 times larger than ours, the parallax would become 1.52 times larger; for Sirius, this would become .38" x 1.52 = .58" of arc.

4. At 10 light years, the sun would be 632,715 A.U. distant, so its angular diameter would be that many times smaller than its present 30 arc minutes (half degree), or .00285 arc seconds.

5. Sirius is about 10,000 K, while the sun is 5,780 K and Betelguese is about 3,000 K. Using the blackbody relation of T^4, we get Sirius emitting $(10,000/5780)^4 = 8.95$ times more energy per unit area than our sun, but the sun emits $(5,780/3,000)^4 = 13.8$ times more energy than Betelguese.

6. Wien's Law tells us the wavelength peaks for blackbodies are inversely related to their temperatures. As the sun's peak is at 4,900 Angstroms, this means the peak for hotter Sirius is at a wavelength 1.73 times shorter, or at 2,832 Angstroms, in the ultraviolet. For cooler Betelguese, the peaks lies 1.92 times longer, or at 9,440 Angstroms, in the infrared.

7. For main sequence stars only, $L = M^3$. Thus for Sirius, $L = (2.1)^3 = 9.3$ times more than the sun.

8. Compared to the sun, Sirius is 10,000/5780 K = 1.73 X hotter than the sun. Using Stefan's Law, we find each surface flux is then $(1.73)^4 = 8.96X$ greater than the sun's. If Sirius is 1.7 times the sun's radius, its surface area is $(1.7)^2 = 2.89$ X greater than the sun's. Hence Sirius is 8.96 x 2.89 = about 26 x more luminous than the sun.

CHAPTER 15

Starbirth and Interstellar Matter

CHAPTER OUTLINE

Central Question: How are stars born out of the material between the stars?

15.1   The Interstellar Medium: Gas

      A. Bright nebulas
      B. Interstellar atoms
      C. 21-cm emission from hydrogen
      D. Clouds and intercloud gas
      E. Interstellar molecules
      F. Molecular clouds

15.2   The Interstellar Medium: Dust
      A. Cosmic dust
      B. Dust and infrared observations
      C. The nature of the interstellar dust
      D. Dust and the formation of molecules
      E. Formation of cosmic dust

15.3   Starbirth: Theoretical Ideas
      A. Collapse models
      B. Protostar formation
      C. Collapse with rotation

15.4   Starbirth: Observational Clues
      A. Signposts for the birth of massive stars
      B. The birth of massive stars
      C. The birth of solar-mass stars
      D. Molecular outflows and starbirth
      E. Planetary systems?

Enrichment Focus 15.1: Emission Nebulas: Forbidden Lines

CHAPTER OVERVIEW

Teacher's Notes

This chapter is a good place to have an extensive slide show to help illustrate the aspects of the interstellar medium described in this chapter. The Trifid Nebula (M-20 in Sagittarius) is particularly interesting as it shows both red emission and blue reflection regions. Most slides of the Pleiades show the nebulosity around the young stars; picture of old open or galactic clusters do not show the gas and dust clouds. Slides will illustrate the marked difference between dark clouds and the surrounding dust-free regions. We especially recommend the Royal Observatory Edinburgh color slides available through the Hansen Planetarium.

Students are often confused by the physical state of the interstellar gas. Be sure that they know the difference between molecules, atoms, and ions. Each form of the gas requires a different observational technique; you might want to use these to sort out the state of the gas.

The detection of dust also represents a problem. The best technique involves polarization, a difficult concept for novice students. Be sure you have them do some experiments with polarizing filters to give them a feel for the phenomenon. You might ask them if they know how polaroid sunglasses work. (The reflection of sunlight off nonmetallic surfaces is polarized, whereas direction sunlight is not. Polaroid sunglasses selectively reduce this kind of glare).

Note that in this chapter serves two purposes: one, to describe the contents of the interstellar medium, and, two, to connect those contents to the evolution of stars by starbirth. These establish the basis for the next three chapters. Note that our view of pre-main sequence stars has undergone a radical change, largely because of X-ray observations. Most (all?) such stars appear to go through a stage of magnetic stellar activity, typified by episodes of flaring and gusty stellar winds, which sometimes appear as bipolar outflows.

15.1 The Interstellar Medium: Gas

We are born from dust...the dust for the interstellar clouds from which new stars and planets are forming even now. With the new technologies of the last decade, we are gaining new insights into the actual process of starbirth.

The essential difference between H II and H I regions is the temperature of the hydrogen. If it is over 3,000 K, the electrons are stripped from the nuclei and glow with the red emission lines so easy to photograph. But if the nebula's temperature is less than 3,000 K, the nebula is crystal clear and transparent, just like neutral hydrogen gas produced in a chemistry laboratory.
The H I regions are mapped via their 21 cm radio emissions, as shown in Figure 15.5.

As discussed in the previous chapter, molecules are fragile things, easily broken up by high temperatures. This is the reason very few molecules have been noted visually, but the list is now well over a hundred known via radio and microwave studies. These are the windows through which we can look at a universe cold enough to produce many of the molecules we have seen in the spectra of comets. Certainly such molecules are present when the solar system formed, and their abundance throughout the star-forming regions of the Milky Way makes many astronomers think the chances of life elsewhere are good, based on this chemistry alone. The discovery of the PAH molecules in the Allan Hills meteorite in August 1996 certainly reinforces the idea of universal abundance and distribution of these complex carbon based molecules.

149

I like to tell my students there is something in the giant molecular clouds for practically everyone. There is formaldehyde to embalm you, ammonia to scrub your toilet, methane to power your furnace, acetylene to power your head light, formic acid for your blue cheese addicts, ethanol for addicts of another sort, vinegar for your salad dressing, and even a form of octane for your car. Many of the critical amino acids for proteins and life itself have now been cataloged as well.

15.2 The Interstellar Medium: Dust

Stress that our own atmosphere, while far denser than the interstellar medium, is still a reasonable model of processes happening out there. For instance, tiny dust motes here scatter best the waves close to their own size; thus our sky is blue. The particles near the Pleiades are similar in size, as the color in figure 12.10 plainly shows. When the sun is setting, its light must pass through a deeper cross-section of the atmosphere, so more light is scattered, and the sun or moon on the horizon appear orange or red; only those waves are long enough to be transmitted.

The appearance of the dark clouds make them seem thick. Not really; consider that clouds of water vapor make up less than 1% of the weight of the atmosphere, yet can darken the sky in storms tremendously. Likewise the dust from volcanoes can darken the skies, even though these dust clouds are not much more substantial than clouds of water droplets.

The dust clouds probably comprise less than 1% of the mass of the galaxy, yet they manage to block over 90% of the Milky Way from our direct visual observations. For this reason, we must use other wavelengths, especially infrared and radio waves, to peer through the dusty veil and map the whole galaxy. For the same reason, we use infrared to detect the first energy radiated out by young stars forming in the darkest, densest regions of the nebulae. Visually the births of stars, like human children, are very private affairs, with the young protostar cloaked in a cocoon of gas and opaque dust. The comparison of Figures 15.14 and 15.15 reveals this distinction well. Of course, energy is not wasted; if the dust absorbs the visible light, the absorbed light in turn heats the dust, until it becomes hot enough to become an infrared source.

Note how closely the interstellar dust mimics the composition of the terrestrial planets. Note that the dust grains include ices like comets, and silicates and iron like the asteroids. These dust grains have cold surfaces, the ideal place to synthesize the complex molecules we find in the giant molecular clouds.

15.3   Starbirth: Theoretical Concepts

No star is born alone; all stars are born in clusters, like the one in the Lagoon Nebula. In the collapse of the nebula into protostars, the first stars to form are the most massive, and hence the most luminous ones. As discussed below, even the sun was once part of such a cluster. As these massive stars explode as supernova, they push out the gas and dust from the central regions of the nebula.

The cluster is never complete. The first stars to collapse will be the most massive, and with their high luminosities, the ones with the shortest stellar lifetimes. They will go supernova in just a few million years, less time than it takes the least massive stars to even start shining.

A human analogy is my mother's family. She was the youngest of twelve children, and before she was born, the two eldest had already passed away. Like her family, no star cluster can ever have a complete "family reunion".

15.4   *Starbirth: Observational Clues*

Note that not only do high mass stars collapse from the nebula and form more quickly, but that they also evolve far more rapidly. A cluster is never "all there"; by the time stars of a single solar mass shed their cocoons and emerge on the main sequence, the most massive members have probably already exploded as supernova. Look at photos of the Rosette Nebula to see how this has already happened, with the center free of gas and dust already.

The role the T-Tauri wind played in the solar system's development is worth more discussion. When the sun's nuclear fires ignited, its radiation drove the light volatiles outward, cleansing the inner solar system of most gas, but leaving the heavier dust particles behind to form the terrestrial planets. In the outer regions, the jovians swept up the outbound gas, and grew huge.

The bi-polar flows illustrated in figure 15.25 will show up time and again where gravitational collapse has produced a disk of rotating material and strong magnetic fields. The T-Tauri wind associated with them goes a long way to explaining why the volatiles were stripped from the inner solar system and pushed farther out. New Hubble photos show this well in the text supplement.

ADDITIONAL RESOURCES

Slides 15.19abc and 15.21 are, like the other slides, unique to this text. Transparencies 15.7, 15.8, 15.16, 15.17, 15.18, and 15.25 also keep your presentations current.

If you haven't already shown it, this is the ideal chapter to show the ninth episode of Cosmos, "The Lives of the Stars." The Orion Nebula zoom is a masterwork of astronomical graphics.
The "Stardust" episode of The Astornomers videos is also appropriate.

SUPPLEMENTARY ARTICLES
1. "Stellar Mass Loss (I)," <u>Astronomy</u>, September 1985, page 78.
2. "Stellar Mass Loss (II)," <u>Astronomy</u>, November 1985, page 94.
3. "The Material between the Stars," <u>Astronomy</u>, June 1988, page 6.
4. "The Births of Sunlike Stars," <u>Astronomy</u>, September 1988, page 22.
5. "Inside Orion's Stellar Nursery," <u>Astronomy</u>, August 1989, page 40.
6. "Voyager into the Interstellar Medium," <u>Astronomy</u>, February 1990, page 42.
7. "The Genesis of Binary Stars," <u>Astronomy</u>, June 1991, page 34.
8. "Star Dust," <u>Astronomy</u>, March 1992, page 46.
9. "Unfolding Mysteries of the Stellar Cycle," <u>Astronomy</u>, May 1992, page 42.
10. "Desperately seeking Jupiters," <u>Astronomy</u>, July 1992, page 42.
11. "Stars Blowing Bubbles," <u>Astronomy</u>, January 1993, page 46.
12. "Clusters and Nebulae of the Summer Milky Way," <u>Astronomy</u>, June 1994, page 62.
13. "The Newest Stars in the Orion Nebula," <u>Astronomy</u>, November 1994, page 40.
14. "Needles in the Cosmic Haystack," <u>Astronomy</u>, September 1995, page 50.
15. "Lone Star Infants," <u>Astronomy</u>, February 1996, page 36.
16. "Seeing the Unseen," <u>Astronomy</u>, March 1996, page 70.
17. "Two New Solar System," <u>Astronomy</u>, April 1996, page 50.
18. "Starmaker," Astronomy, <u>Astronomy</u>, July 1996, page 52.
19. "Plunge into the Lagoon," <u>Astronomy</u>, July 1996, page 36.
20. "The Excesses of Young Stars," <u>Astronomy</u>, September 1996, page 36.

ADDITIONAL DISCUSSION TOPICS

The Brothers and Sisters of the Sun

As the sun was not born alone, why is it alone in space now? When the cluster forms, the first stars form in the center and are the most massive and luminous. They are also the ones which evolve the fastest; some of them may have become supernovae before the sun even reached the main sequence. In fact, it may be in part the shock waves from these supernovae that speed the gravitational collapse of the rest of the nebula into the smaller stars in the outer regions of the cluster. As the sun was one of the later stars to form in the outer regions of the cluster, it would not have been as strongly bound to the cluster as the more massive stars closer to the center; like an Oort cloud comet, it would be more likely to escape the gravity of the cluster if the cluster passed close to another cluster. As the Galactic Year is about 250 million earth years, our cluster probably has circled the galaxy about 20-25 times by now, with plenty of encounters to loosen up the cluster and allow many (if not all) of its members to escape into space. Such open clusters are often referred to as loose clusters for this reason.

Few known open clusters are even known to be as old as the solar system--most may fall apart in a billion years or less. They are also known as galactic clusters, for they stay in the galactic plane. This in term means that they orbit in the thick of things, where close encounters with other comparable clusters are likely.

Given these constraints, does our home cluster even still exist? Probably not. M-67 in Cancer is one of very few open clusters which appear to be about five billion years old. In such an aging cluster, the bright blue stars which make the Pleiades so beautiful are long gone; instead its brightest members will be swollen red and orange giants like Arcturus, perhaps a 100x as luminous as the sun. The main sequence stars still remaining will be classes F5 and cooler; any star more massive and luminous than this will have already used up its main sequence supply of hydrogen. If it is our parent cluster, then its motion around the Milky Way's center should be similar to the sun's present orbit, almost circular and inclined about 5 degrees to the galactic plane. Of course, after 20 or so trips, by now our sun could have drifted to the other side of the galaxy. But just like DNA matching, we still should be able to find G type stars in this cluster with essentially the same proportions of elements that appear in the sun's spectrum. Our star and this cluster were formed from the same nebula.

Even if the cluster no longer exists, some of the less massive members should still be around, perhaps in our neighborhood. As mentioned above, they will not be the bright blue stars that dominate the night sky; such massive stars will not last nearly long enough to still be visible. But perhaps some of our cooler neighbors, stars like alpha Centauri or epsilon Eridani or Tau Ceti, might be nearby because they have always been close to us. Again, close spectral examination, perhaps with the Hubble Space Telescope, might be a way to establish such kinship.

Another clue to their relationship to us will be their motion through space. If they have space velocities similar in speed and direction to the sun's, then they might well have come from the same cluster and still be travelling in parallel paths to the sun's, much as cometary debris still moves in the old comet's orbit and creates meteor showers long after the nucleus disintegrates.

If such stars are found, and they are indeed our brothers, the same age and chemistry as our sun, then planetary systems around our sibling stars might be abodes of life worth talking to. Let the search begin.

Stellar Mass Ranges

While stars vary tremendously in luminosity, with the brightest about a million times more luminous than the sun, and the faintest 10,000x fainter, the actual range of masses is far less. Why are there no stars yet found of over 60 solar masses, nor any less massive than about 8% of the sun's mass?

The answer for the top end is uncertain; it may be that such a protostar would heat up so fast as to drive off much of the material around it before it could grow any larger, or as discussed with angular momentum, it may become unstable and form a binary or multiple system, rather than just one huge star. A third possibility discussed in Cosmos 109 is that somewhat heavier stars might form, but they would evolve rapidly and blow up as supernova before they ever had a chance to emerge from their protostellar cocoon.

At the bottom end of the mass range, the answer is far easier; below about .08 solar masses, the gravitational energy is not sufficient to allow the proton-proton cycle to start fusion of hydrogen into helium. While such "brown dwarfs" may even shine for a few million years due to the energy of gravitational collapse, they will never evolve into main sequence stars and fade rather soon; Jupiter is in this sense a now cool brown dwarf, a star that failed. Such objects may in fact be abundant; some speculate they represent the majority of the mass of the universe, but finding them for certain has been difficult. Again, new technologies, particularly with infrared telescopes, hold promise of finding many of these objects soon.

DEMONSTRATION AND OBSERVATIONS

Observing Emission and Reflection Nebulae

Several stellar nurseries are within range of amateur scopes or even binoculars. The Orion Nebula (M-42) is the highlight of the winter sky; with larger scopes, it is perhaps after Saturn the next most beautiful object in the sky. While visible faintly with the naked eye under good conditions, binoculars help a great deal and reveal it to be two separate Messier objects. M-42 is the southern bright emission nebula, while just above it M-43 is a bluish reflection nebula. If M-43 were not on the edge of M-42 and thus overshadowed, it would also be rated as one of the sky's show objects.

Another reflection nebula envelopes the famed Pleiades cluster. With a 4" reflector at 15x on very clear evenings, the entire cluster in enveloped in haze. I tell my students that the Pleiades is a cluster so young it is still running around in its diapers, due to the cotton batting appearance of this nebula in photos.

The Rosette Nebula photographs well, but doesn't show up visually; look for the cluster lying in the center of it about 10 degrees east of Betelguese. Likewise the famed Horsehead shows up well in long exposure photos of Orion's belt (it lies just south of the bottom star) but I have never glimpsed it in anything smaller than a 10" telescope. The reason is that the hydrogen alpha line lies near the extreme red end of our retinal sensitivity, so our eyes do not respond as well to it as do many photographic emulsions. To me, the center of the Orion Nebula looks more greenish than red; our eyes pick up the emission line from ionized oxygen better than they do the hydrogen alpha line, even though the latter shows up much brighter on photos. Look for advertisements by Lumicon in astronomy magazines for deep sky filters that block light pollution and make these types of nebulae stand out much more vividly.

The spring Milky Way lies south of our horizon, but more southern observers can see the finest emission nebula of all, eta Carinae. The superstar in its center is variable, and a potential supernova. 180 years ago it rose in brilliance to become second only to Sirius in brightness, but now it has faded temporarily down to sixth magnitude.

Summer skies contain several notable nebula besides the bright Lagoon Nebula. M-8, like the Orion Nebula, is visible with the naked eye. Its name comes from the dark nebula lying in front of the bright red H II region, appearing to split it with a dark embayment, the lagoon. Note the young cluster just to the east of the nebula itself.

Just above M-8 is the smaller but more colorful Trifid Nebula, M-20. Again dark dust lanes appear to split the nebula. Other good emission nebula for binoculars in the Sagittarius region include M-16 and M-17; find them with your star atlas.

Farther north, in the same binocular field as Deneb, is the famed North American Nebula, an H II region that can be glimpsed under very dark sky conditions. It shows up very readily in time exposures of five minutes or longer.

Remember that for every region of interstellar gas hot enough to glow brightly as an emission nebula, there are huge clouds of cold, neutral hydrogen, or H-I regions. These are transparent, and we must use 21 cm radio waves to map their distribution. But if there is a high enough concentration of dust in the cold clouds, it will redden the light of stars behind it, or if dense enough, block their light completely. The Horsehead itself is actually such a dark nebula; we see its silhouette only because it lies in front of a faint H II region. The most striking dark nebula in the sky is the dusk lane called the Great Rift. It splits the visible Milky Way into two lanes in southern Cygnus and down through Aquila.

Other smaller dark nebula can be traced out with binoculars due to the absence of stars in that direction along the Milky Way. Next to the southern cross is the aptly named Coalsack Nebula of the southern sky. It is almost the same size as the Cross, and lies just below and to the east of that famed southern constellation.

ANSWERS TO STUDY EXERCISES

1. H I is observed at 21-cm wavelength with radio telescopes. H-II regions glow bright red from hydrogen alpha emission and many are seen visually. The coronal interstellar gas is quite hot, and like the solar corona, is best observed with ultraviolet or X-ray telescopes. Molecules must form at far colder temperatures, so their molecular lines are often in the millimeter bands in the radio or microwave spectrum.

2. Interstellar dust will both dim and redden starlight passing through it. As the dust scatters blue light more (such as the nebulosity surrounding the Pleaides), the stars will seem redder in color than their spectral lines reveal. If the dust is thick enough, the stars may be so dimmed that "dark nebula" completely block the light of stars behind them (the Horsehead Nebula in Orion is the most famous example). Also, the dust grains can polarize the starlight in transit. Light radiated by the star's photosphere is naturally unpolarized; if magnetic fields align the dust grains in space, the light reaching us may be polarized by selective absorption by these dust grains.

3. The most likely components of the dust grains are the common substances in the solar nebula. Some dust grains are of almost pure carbon. Others may have iron mixed in with silicates, allowing them to be aligned by magnetic fields. Other grains may have carbon or rocky cores surrounded by coatings of various ices; exposure to ultraviolet light may degrade some of these carbon and hydrogen rich compounds into tarry hydrocarbons, such as in carbonaceous chondrites.

4. (a) Millimeter observations of giant molecular clouds close to H II regions.
(b) Infrared sources embedded in the giant molecular clouds.
(c) Small clusters of hot, young stars detected by radio or infrared observations.
(d) Expansion of hot gas around young stars.

5. Conversion of gravitational potential energy into other forms, chiefly light and heat, as the protostar contracts and heats up.

6. The star forms inside a dusty coccoon, and the dust around it absorbs its light, blocking out view. As the star heats up to the T-Tauri stage, it blows away this coccoon. Eventually the dust and gas thin out enough that the star's light shines through, usually when the star has almost reached the main sequence.

7. A massive star is so hot that its ultraviolet radiation ionizes much of the surrounding gas, casuing the hydrogen to glow bright red as an H II region. The radiation pressure will eventually blow the gas away from the star entirely, as is shown by the central region of the Rosette Nebula.

8. We find infrared sources that have the characteristics of solar-mass protostars embedded in some dark clouds.

9. Doppler shift observations of the molecular clouds, which show both blue and red shifts. These often take the form of bi-polar outflows, perpendicular to the dusty disks forming around the protostar's equator.

10. Because stars are nuclear reactors that shine by consuming themselves, the bright blue stars of types O and B, having very short lifespans, must constantly be replaced as the evolved ones blow up as supernovae.

11. The most likely mechanism is the propogation of a shock wave through the cloud.

12. Stellar winds from cool giant and supergiant stars can condense to form some kinds of interstellar grains; some red giants literally blow "smoke screens" of carbon dust around the star, vanishing for months at a time.

ANSWERS TO PROBLEMS & ACTIVITIES

1. Its peak wavelength is (Wien's Law) at 2.9 x 10-3 / 600K = 5 micrometers. Its luminosity is 1000 x 3.86 x 10^{26} J/sec = 4 x 10^{29} watts. The radius can be found from L = 4 pi R^2 x Flux, where the flux = 5.67 x 10^{-8} (600 K)^4 = 7,400 watts per square meter. From this, we get R^2 = (4 x 10^{29})/9.3 x 10^{7} = 4.3 x 10^{24} square meters, so R = 2.1 x 1012 meters or 2.1 billion km; about 13 AU in radius, well out beyond the orbit of Saturn!

2. Using the ideal gas law, P = nkT, with estimates from both molecular clouds and H II regions, we find for the molecular cloud that P = (1.38 x 10^{-28})(10^{12})(10) = 1.4 x 1^{-15} atm. For the H II region, we find P = 1.38 x 10^{-28} (10^9)(10,000) = 1.4 x 10^{-15} atm., a very similar pressure.

3. One AU = 150 million km, so at 100 km/sec, it would take the outflow 1,500,000 seconds, or 17.4 days to move this far out.

4. Dust-absorbing light is heated enough to give off the infrared radiation detected by the IRAS satellite in 1983, evidence that protosplanets are forming around this young star.

5. Water has a density of 1000 kg/m^3, so if you weighed 100 kg (about 220 pounds), you would have about .1 cubic meter of water in your body. The water molecule weighs about 18 amu, or about 3 x 10^{-26} kg; this means your body contains about 100 kg / 3 x 10^{-30} kg/molecule, or about 3.3 x 10^{27} water molecules. Given a number density in a molecular cloud of 10^{12} particles per cubic meter, your water molecules could spread out to cover a volume of 3.3 x 10^{27} /10^{12} =   3.3 x 10^{15} cubic meters, or about 3.3 x 10^{16} times less dense than your body.

CHAPTER 16

Star Lives

CHAPTER OUTLINE

Central Question: As the stars go through their normal lives, how do their physical properties change?

16.7   Observational Evidence for Stellar Evolution
        A. Stars on groups
        B. Globular clusters
        C. Stellar populations
        D. Comparison with the H-R diagrams of clusters
        E. Variable stars
        F. Central stars of planetary nebulas

16.8   The Synthesis of Elements in Stars
        A. Nucleosynthesis in red giant stars

Enrichment Focus 16.1: Degenerate Gases

CHAPTER OVERVIEW

Teacher's Notes

The picture of stellar evolution marks one of the triumphs of modern astrophysicists and provides the best example of the interdependence of the theoretical models and observations. The magnitude of the problem of determining a star's structure and evolution can be put in perspective by again considering an analogy to human beings, as in the beginning of the chapter. Image the problem of finding out all there is to know about the human animal: structure, life cycles, behavior, etc. Then add the constraints that you cannot touch any one directly, that only one is reasonably close to you, and finally that you have only one minute in which to make observations (one minute out of a human lifetime amounts to the same fraction as the period of modern observations of stars compared with the average stellar lifetime). You might pose this problem to students and ask how they would approach the problem.

The same basic techniques and assumptions used by astronomers should be suggested, for instance: observe a large number of specimens; observe in as many different ways as possible; assume that people age; assume that basic structures and physical laws apply to all specimens. Admittedly the analogy isn't exact; stars are more predictable that people. But you might find that an operational procedure suggested by the students will be remarkably similar to those followed during the past hundred years or so by astronomers.

It might not be readily apparent to students that stars of greater mass have shorter lifetimes. Students may be aware, from chemistry courses perhaps, that higher temperatures make particle reactions go faster, and may be able to see that more massive stars will have higher temperatures in their cores. But on a qualitative level, the fact that there's more mass of more fuel could nullify the increased rate. The key factor is the reaction rates for the proton-proton chain and for the carbon cycle are very sensitive to temperature. For the proton chain, the reaction rates depend approximately on temperature as T^3 or T^4; for the carbon cycle, the dependence is much stronger, approximately T^{14}. The central temperature of star is roughly proportional to the mass. So doubling a star's mass will cause proton-proton chair reactions to go about sixteen times faster, and carbon cycle burning will go 16,000 times faster. Order-of-magnitude examples using the dependencies will correspond to lifetimes given in the text.

The warning given about motion on the HR diagram in the text should be taken seriously. Students having problems with abstract concepts or reasoning may confuse a star's position on an HR diagram with its position in space. The language used by astronomers doesn't help matters. We speak of stars moving "along a path" or "to the right of the main sequence." Many students picture the star as moving through space along that path, or in that direction. Analogies using everyday examples may help in this case. For example, people may be classified by height and weight, as shown in the text. You might record these two characteristics for each student, then plot points on the graph for each variable. you would find most students lie along a line, analogous to the main sequence on an HR diagram. Students can guess how the graph would look if one person's height and weight were measured and points plotted over five-year intervals throughout his or her lifetime. The line generated would be an evolutionary path, and NOT represent the person's actual motion through space! For more details and examples, see "Effective Astronomy Teaching: Intellectual Development and its Implications," by Dennis Schatz and Anton Lawson, Mercury, July/August 1976.

When wrapping up this chapter, you might point our that a scheme might be self-consistent and still be wrong. One danger in having answers to all questions is that we stop being skeptical of those answers. Recall the solar neutrino problem from Chapter 13. Some people believed the experiments to be invalid before they considered the problem might be in our understanding of the sun's interior or the properties of the neutrino; now several other detectors worldwide have all proved the neutrino deficiency is real, and we have theoretical problems.

16.1 Stellar Evolution and the Hertzsprung-Russell Diagram

Stars are nuclear furnaces that shine by consuming themselves. The most massive stars are gluttons that rapidly consume their hydrogen, then as giants turn on the afterburners, producing still heavier elements in an ultimately vain attempt to keep fusing and producing energy. Through their lives, the stars will take on vastly different appearances and dimensions, ranging from sun-sized main sequence to gigantic supergiants to planet sized white dwarfs or even city sized neutron stars or black holes. In determining the star's fate, it is its mass that predestines the course of its evolution from birth to death.

Stress again that main sequence stars are so common because there is a lot more hydrogen in the known universe than all other elements combined, so it makes sense than at any given moment, most stars will be using it for their nuclear fuel. As to why mass is so important, point out to your students that more mass means greater gravitational pressure as the star forms, leading to higher internal temperatures and a larger core capable to fusing the hydrogen into helium at a faster pace. The quicker hydrogen turns into helium, the more energy is released and the more luminous the star becomes.

16.2 Stellar Anatomy

Every star lies on the edge of two disastrous fates. If gravity won, the star would collapse to become a black hole. If radiation were to win completely, the star would be blown apart, like a type I supernova. But as long as sufficient fuel reserves to continue the fusion processes exist inside the star, such catastrophic changes are averted and the star's evolution is fairly gentle.

The hot region near the star's core is transparent to gamma rays, so they move freely through this radiative zone. But closer to the photosphere, as the cooler gases become opaque to radiation, their atoms absorb the gamma rays. Any atom which absorbs a gamma ray has gained a lot of kinetic energy--it is now hotter, moves faster, and collides with its neighbors more frequently. Each of these collisions imparts some of the gamma ray's original energy to the atom which was just hit. It in turn strikes its neighbors, etc. By the time the photon of gamma ray energy has passed through the convective zone and is radiated away by the granules into space, it has been so subdivided that it make about six million photons of yellow light. Consider the wavelengths; a typical gamma ray might have a wavelength of about .001 A, but the photons of yellow light are around 6,000 A, or 6,000,000 longer than the gamma ray. The 3 L law reminds us that longer waves mean lower frequency and less energy per photon. That is fine with me; I much prefer visible light to gamma rays from the sun.

16.3 Star Models

In the giant stage, the core must become hotter to turn on the afterburners, and fuse the heavier elements. Why then do giants typically appear redder, not bluer than they did as main sequence stars? The answer is that the extra radiative pressure pushes out hard enough to partially overcome gravity and disrupt the hydrostatic equilibrium. The part of the star we see, the photosphere, is distended and swollen by the extra radiative pressure. As you know from the chill of your underarm spray deodorant, expansion is a cooling process; the gases in the outer layers of the expanding photosphere cool off, causing the star to appear to evolve to the upper right off the main sequence during this giant phase expansion.

16.4 Energy Generation and Chemical Compositions

It is believed the "break point" for CNO vs. P-P is about class F; stars hotter than class F probably get more energy by using carbon as a catalyst. But as this works best at higher internal temperatures, stars G class and cooler have to get along chiefly with the proton-proton cycle. Remember both yield exactly the same products and energies; the catalytic action of the carbon merely allows the star to get its energy released faster, making them more luminous.

16.5 The Theoretical Evolution of a 1-Solar-Mass Star

Not everyone agrees on just how big and luminous the sun will become in its red giant stage; Zeilik says 1,000x its present luminosity, at half its present temperature, and large enough to swallow up the earth as well as Mercury and Venus. But others say 100x its present luminosity is more realistic, meaning the sun will not even get to devour Mercury. In any case, it will be luminous enough to certain make the earth uninhabitable, so our descendants must consider moving on.

Note that as more and more types of fusion reactions occur in the star's interior, its evolutionary path gets more complicated. These giant stars are now long period variables, with their luminosity changing considerably in just a few years.

Planetary nebulae can be also thought of as the funeral wreaths of dying solar mass stars. In the case of the beautiful Helix Nebula, it must have been a nice service. May our own solar system go so gently into the dark cosmic night. The newest Hubble Space Telescope photos of planetaries such as the Helix, Cat's Eye, and Egg Nebula show that the shedding of mass is complex and very intricate and beautiful. Be sure your students note the spectacular colors and details.

160

Since the text states that even the earth will be swallowed up by the red giant sun, it marks the tombstone of the atoms which are today us and our planet as well. But since those atoms are being shed back into space, there is also in the expanding planetary nebula the promise of rebirth, as the atoms mix into the interstellar medium, to be recycled through a new generation of stars and planets billions of years from now; nitrogen in particular seems to be made in planetaries.

I doubt there are any black dwarf stars yet. White dwarfs contract and cool very slowly, and even the oldest white dwarf in the cosmos may still be hotter than the sun. In a sense, the white dwarf contraction may represent an even longer span of time than the star's stable main sequence lifespan...perhaps a prolonged retirement is a good analogy.

We think that only stars of .4 solar masses up to 2 solar masses will swell to red giants and then pass through the planetary nebula phase. In the range of .08 to .4 solar masses, hydrogen burning proton-proton reactions will occur, but there will not be sufficient pressure to get the core up to 100 million K, where triple alpha can begin to swell the star outward in the red giant phase. Theory leads us to expect that these stars will collapse directly from the main sequence to white dwarf size, but since these misers are going to stretch their hydrogen supply out over trillions of years, don't wait up for it to happen.

Again, if there is less than eight percent of the sun's mass present, the object will never begin hydrogen fusion and turn into a main sequence star. Brown dwarfs are now being found all around us; see the "Needles in the Cosmic Haystack" article on page 50 in September 1995 Astronomy. Since the less massive stars are far more abundant than stars like the sun and heavier, it is possible that very abundant brown dwarfs represent the majority of mass in the cosmos. The search continues.

16.6   Theoretical Evolution of Massive Stars

Zeilik notes, as I do below in discussing planetary nebulae, that the star may pass though more than one red giant phase, then expel material as a planetary nebula, then swell back up again. For instance, Betelguese in Orion's shoulder is a notable variable that is now surrounded by a tear shaped shell of gas and dust its strong stellar winds have shed into space in the last few thousand years. Whether this mass loss will be sufficient to prevent the complete core collapse into a supernova is still uncertain. Dr. Virginia Trimble of MIT, one of the world's greatest experts on supernova, gave me personal odds of 1 in 100 that I would live to see Betelguese blow up; if it is only 300 light years away, this event would be brighter than the full moon and probably dangerous to look a with the naked eye. What a fireworks display that would be.

16.7   Observational Evidence for Stellar Evolution

As mentioned already in Chapter 15, a cluster is never all there; the most massive stars will evolve so rapidly that they explode as supernova before the lower mass main sequence stars emerge from their cocoons. Thus a young open cluster, like the Pleiades, is dominated by bright blue, very luminous but short-lived massive main sequence stars. An older cluster, like the one which gave birth to our sun, or even older globulars, have long since lost such massive stars to black holes or neutron stars; the brightest remaining visible members have now evolved into red giants, turning off the main sequence to the upper right. As shown in Figure 16.15, the lower down the main sequence this turnoff point, the older the cluster. For instance, NGC 188 might be a candidate for our home cluster, now five billion years old. Perhaps the spectroscope can tell us of kinship ties.

Why do we concentrate on the stars in clusters, rather than random stars in general, to study evolution? Well, all the stars in the cluster were, as just noted, born at about the same time. If the more massive ones have evolved off the main sequence, it is because they are evolving faster. They also were born of the same nebula; if there are now chemical differences among the stars in the cluster, those differences have arisen since all the stars were born, in the interiors where these new elements are synthesized by the more massive stars evolving faster.

In considering stellar populations, remind the students that Population II came FIRST; the numbers were unfortunately assigned by Walter Baade before any evolutionary implications were realized. Since for us our sun is a number one star, it and its fellows nearby in the disk are also population I as well. As the galaxy ages, subsequent generations of supernova will continuously enrich the interstellar medium with more and more heavy elements. Thus the oldest population II stars are almost pure hydrogen and helium, the gifts of the Big Bang itself. Only about .1% of their mass is heavier elements, probably contaminants from supernova near them showering them with a few heavier elements. Our star is about 1 % metals (anything heavier than helium), while the new stars now forming in the Orion Nebula are about 4 % metals. As the universe ages, there will be less and less hydrogen, and more heavy stuff.

16.8   The Synthesis of Elements in Stars

The sun will probably never produce anything heavier than carbon in its core. The production of heavier elements takes even higher temperatures, thus even higher masses than the sun. Note in table 16.3 that we are heading toward iron production at over three billion degrees K in the star's core. As iron is the most stably bonded nucleus, there is no way to get radiative energy from it to continue the normal radiative life of the star, setting the stage for type II supernovae and the recycling of the new elements produced outward into the interstellar medium in the next chapter. This is notable in the periodic table, as carbon and oxygen are considerably more common than even the even-numbered elements which follow; far more stars can synthesize carbon and oxygen than can go all the way to making silicon and sulfur, for instance.

ADDITIONAL RESOURCES

The slide 16.11 of planetary nebulas and transparencies 16.2, 16.3, 16.4, 16.5, 16.10, 16.12, 16.15, 16.18, 16.20, and 16.21 add color and graphics to your class.

The fifth episode of the Astronomers, "The Stars," is widely available from video stores and focuses well on the processes of stellar evolution, particularly star death and Supernova 1987a. It fits well with either this chapter or the next, as does "Stardust" from the Astronomers series.

SUPPLEMENTARY ARTICLES
1. "Stellar Interiors," Astronomy, December 1984, page 66.
2. "Observing Variable Stars," Astronomy, January 1985, page 74.
3. "The Origins of the Elements," Astronomy, August 1988, page 18.
4. "Unfolding Mysteries of the Stellar Cycle," Astronomy, May 1992, page 42.
5. "Giving Birth to Supernovae," Astronomy, December 1992, page 46.
6. "Stars Blowing Bubbles," Astronomy, January 1993, page 46.
7. "Seeing a Star's Surface," Astronomy, October 1993, page 34.

8. "Hypergiants," <u>Astronomy</u>, March 1994, page 32.
9. "Ashes to Ashes," <u>Astronomy</u>, May 1994, page 40.
10. "White Dwarfs Confront the Universe," <u>Astronomy</u>, May 1996, page 42.
11. "Unwinding the Helix," <u>Astronomy</u>, July 1996, page 44.
12. "Faint Balls of Fire," <u>Astronomy</u>, September 1996, page 74.

ADDITIONAL DISCUSSION TOPICS

The Intelligent Mayfly

While mayfly nymphs live several years underwater, their adult life is spent in a brief day of flying around, looking for a mate. Let us imagine an intelligent mayfly, who having passed on his genes, decides to spend the remaining hours of his existence finding out about the world upstairs. As there are so many of us, he might choose humans to focus his investigation upon.

Were he to flit into your classroom, he would probably find an uninteresting lot. Most of your students are similar in size and appearance, although he might perceive gender differences. He would find a mall much more interesting, for there he would observe people of all ages, from babies to retirees. Still, they would probably be mostly in good health and not give any clue to major life changes as he watched them.

If very lucky, our mayfly would flit inside a general hospital. In the maternity ward he might see babies being born; in critical care, he might even witness a death. Taking extensive notes, and feeling his own death eminent, our mayfly flits back to the pond, then drops off his notes for the next generation of his prodigy to assimilate in their years of study at the nymph school on the stream's bottom. Thus mayflies learn about the human life cycle.

The analogy is, of course, to stars and our own human life span. The time spans are carefully chosen. The day's life of adulthood for the mayfly represents a span 25,000 times less than that of an average human being; our proverbial three score and ten in turn is about 25,000x shorter than the two million year life of a massive blue star, such as Deneb or Rigel; yet these are the shortest lived and fastest evolving of all stars. Any single human generation is unlikely to see major changes in a given star, but with records extending for thousands of years, we are now seeing patterns as young stars emerge from the Orion Nebula, and aging stars die as supernovae, such as in 1987. As Newton so aptly put it, "If I have seen farther than those who came before me, it is because I have stood on the shoulders of giants."

Why Odd Elements are Rare

Look at the periodic table (you need one posted in every astronomy classroom). Ask the students once they get to carbon, which are more abundant, the odd or even-numbered elements. You will find that, without exception, even-numbered elements are always more abundant than their odd-numbered neighbors on either side. It is the triple alpha process which allows us to understand why this occurs.

In both the P-P and CNO cycles, we were adding protons together, turning four of them into very stable He 4. As the main sequence star ages, it builds up more and more He 4 in its core; unless it comes up with something to do with this helium ash, its nuclear fusion will cease, and the core collapse to a white dwarf (exactly what theory leads us to expect if the star is less than .4 solar masses). So what to do with the helium ash?

163

First, let's try to keep adding on more protons; we still have plenty of hydrogen around to provide these nuclei. He 4 + H 1 = Li 5, so lithium should be the third most abundant element in creation, right? No way; Li 5 is very radioactive, and immediately decays back into the helium and proton you tried to fuse. In our stepladder of elemental evolution, we have found a broken rung. Think of how different the composition of the universe would be, were the Lithium 5 isotope stable, and the third most common element in creation.

Yet there is lithium in the universe, so where did it come from? The stable isotope, Lithium 7, requires two extra neutrons be added to a Li 5 to stabilize it BEFORE it has time to decay. But since neutrons are radioactive, this does not happen often, and lithium is rare.

Well, we have plenty of He 4 around, so let's just fuse two helium atoms (or alpha particles, in radioactive jargon) together. Thus He 4 + He 4 = Be 8, so beryllium must be the third most abundant atom around, right? Wrong again; Be 8 is also radioactive, and usually decays back into the two alpha particles you started with. Sounds like the universe is going to be a rather dull place, made of only hydrogen and helium.

But Be 8 has a saving grace; it is not as radioactive as Li 5, and at 100 million K, there is a statistically good chance a THIRD (hence triple alpha) helium will strike the Be 8 nucleus before it decays. Now we are getting someplace; Be 8 + He 4 = C 12, which **IS** the third most abundant element. And once this hurdle is made, at higher temperatures we can keep on adding more and more helium nuclei to synthesize heavier and heavier elements. Thus four heliums yield an O 16 nucleus, and seven He 4 makes Si 28, et c. As helium contains two protons, and it is the main fuel left in the core, many more even numbered nuclei will be made in these reactions. Of course, there is some residual hydrogen around, so odd numbered nuclei are also made, but not as frequently. These fusion reactions will cease to be energetic once the most stably bonded nuclei, iron, is made, setting the stage for the type II supernova event.

Electron Degeneracy

To visualize electron degeneracy, consider that the white dwarf core is about the size of earth, but contains as much mass as the sun. Since we earlier found that you could pack a million earth volumes into the sun, this means, as Zeilik notes, we have a density about a million times greater than that of normal matter around us, for the collapse has packed the mass of the sun into the volume of earth. This is achieved by utterly crushing the normal orbital structure of the electrons. The atom was chiefly empty space, occupied by the low mass electrons in their quantized orbitals; in the Pauli Exclusion Principle, we used this packing of the electron orbitals to explain the Periodic Table. But gravity disregards Pauli's nice set of rule for proper electron behavior--the degenerate matter still contains atomic nuclei, but the electrons move freely in a dense sea around the nuclei, now packed a million times more closely than in normal matter.

Life and Stellar Populations

Our solar system is only five billion years old. Studies of the H-R diagrams of clusters reveal some that could be three times older. Why then have not intelligent species from those stars already taken over the galaxy? In Hollywood terms, why hasn't "Independence Day" already happened long ago?

164

There possibly are no such species. Our life is vitally dependent on the chemical complexities of the carbon atom and the water molecule. While these are abundant in the arms of the galaxy now, that was not true when the first generation of population II stars condensed. They contained only the hydrogen and helium initially produced by the Big Bang.

The metals which make up our terrestrial planet were not made in the Big Bang but synthesized later in the cores of giant stars. The only kind of planet a population II star could have orbiting it would be a gas giant like Jupiter. Organically, you can't do much with just hydrogen and helium, so even though we have beamed messages to a globular cluster (M-13 in Hercules). I, for one, doubt we will ever receive a reply.

Even if we are not alone, it could well be that ours is one of the first solar systems to be formed with enough heavy elements to make terrestrial planets like the earth possible. The new planetary systems found around nearby stars are all stars younger than the sun, with less time for life to evolve as much. Again, that might be why ET hasn't come calling.

DEMONSTRATIONS AND OBSERVATIONS

Stellar Evolution and the Naked Eye Census

Look up on a clear winter evening. What color are most of the stars you observe? Blue, you say. Then what temperature are they? Hot, of course. At what stage of their evolution will most of them be, then? Massive main sequence stars, not old enough to have consumed most of their hydrogen, you think. Very good. But look more closely; do you see any notable yellow, orange, or reddish stars? Are they too main sequence, but less massive than the sun?

Not likely. Of the six thousand stars your naked eye might glimpse, only six are in fact less luminous than the sun. Yet of the 40 closest stars, the sun is the fourth most luminous. The vast majority of stars, in our neighborhood and probably throughout the galaxy and even the universe, are in fact main sequence stars smaller, cooler, dimmer, less massive, and far longer lived than even our G type sun. But because of their low luminosity, they are so faint, even if like Barnard's star, only six light years distant, that it will require at least a pair of binoculars to see them. None of the bright reddish stars is anywhere near the main sequence.

As Zeilik has already pointed out with the Antares double comparison, that we see such cooler stars at all is due to their immense sizes. The distant supergiant Antares is probably about 300x larger than the sun, and still more luminous Betelguese is around 500-900x larger than our main sequence stars (it again depends on whose distance estimates you believe). Arcturus and Aldeberan are closer giants, perhaps 30-40x larger than the sun. Yellow Capella is about 10x the sun's diameter. Such colorful stars are rare because the giant stage through which they are passing represents a small fraction, perhaps about 10% of the star's main sequence life span. Thus at any given moment only 1 star in 10 will be swollen up as a giant, about the same ratio of blue-to-red stars you find among the brightest naked eye stars overhead.

I like to compare the red giant stage to a terminal disease, with perhaps 10% of the star's life spent in an unstable, constantly changing downward spiral, such like prolonged lung cancer. If you have had to care for such a relative, then you may understand some of the fluctuations at the end of both human and stellar lifespans.

Observing Variable Stars

Norton's Star Atlas has a nice listing of notable variable stars with each of its sky maps. Sky and Telescope publishes a monthly set of variable stars predicted to reach maxima that month. Burnham's guide lists exceptional variables in each constellation, and catalogs the misbehavior of some of the more fascinating ones. Many amateurs with even binoculars or small telescopes are members of the AAVSO (American Association of Variable Star Observers), and do much meaningful research by keeping up with thousands of often unpredictable variable stars all over the heavens. There are not nearly enough professional astronomers to do this, and you can never tell when one of these highly evolved stars will do something really spectacular.

The AAVSO provides its observers with star charts to help find the variable, and with comparison stars in the same binocular or telescope field of known magnitude. Once the observer gets used to making comparisons of brightness (often by defocusing the images to accentuate differences in brightness), he can usually give the apparent magnitude of the variable on a given night to an accuracy of .1 magnitude.

Hundreds of such observations from many observers worldwide are put together by researchers to plot the star's variability curve; in some cases, set pulsations periods of days, weeks, months, or even years emerge, while in others, the fluctuations appear completely random. Again, this is an area where amateurs can play a most valuable role; for more information, write the AAVSO at 25 Birch St., Cambridge, MA 02138, or call (617) 354-0484.

Observing Planetary Nebulae

Observing stellar strip teases turns me on. Watching a star shed almost all of its photospheric veil of gases, to stand revealed as a naked white dwarf fascinates me. Since my teen age years I have observed over a hundred different planetary nebula, and each had its individual charm to me. Most are faint, and their real beauty best shown by Hubble photos like those in the supplement, or tiny, and not recommended for class observation, but a few are bright enough to be admired by everyone.

The Ring Nebula of Lyra, M-57, is perhaps the easiest to find. Look midway between the two southernmost stars in the parallelogram of Lyra with a 60mm refractor at about 30x; it will be a faint oval disk. With bigger scopes the darker center stands out well in contrast, giving it the appearance of a smoke ring. But while the central star stands out nicely in the photo, do not expect to see it visually with anything less than an 18" telescope. This white dwarf is very hot, about 200,000 K according to recent satellite data. At that high temperature, Wien's Law tells us that practically all its energy will be ultraviolet radiation, not visible with the naked eye. This ultraviolet in turn is absorbed by the expanding shell of gas, which is excited and glows with the bright emission lines that color the shell of gas. Note red is on the outside; hydrogen is the lightest element, and will travel fastest in the expulsion. Inside heavier oxygen glows green, nitrogen blue, and sulfur yellow with their brightest emission lines. While the spectrum of most stars at first glance looks like a continuum, a diffraction grating in front of the eyepiece now shows the nebula as several colorful ghost images, each the emission line of an important element. In keeping with the stellar strip tease analogy, M-57 is surrounded by several cold shells left behind by previous events, much like Salome's Dance of the Seven Veils. Numerous other planetaries also reveal such shells in radio telescopes, so it is possible a white dwarf may temporarily rekindle its red giant fires in a process not yet observed.

Just south of the Ring, the Dumbbell Nebula of Vulpecula, M-27, is brighter and easily found with binoculars just north of alpha Sagittae. Its structure is not nearly as circular as the more famed Ring Nebula; I expect some kind of bi-polar outflow accounts for its shape. It is a fascinating object in larger scopes, with a good bit of structure notable with 10" or larger.

In the autumn sky, M-76 in Andromeda is a smaller version of the Dumbbell, but in the hard area to star hop to; check you star chart well. Father south, NGC 7932, the Helix Nebula (figure 16.11) of southern Aquarius, is the largest and closest of these objects. It is half as large as the disk of the full moon and under very large, clear sky conditions visible with binoculars. But the large disk is of low contrast, so a telescopic observer using over about 30x could easily sweep right past it without noticing the faint, circular disk. This nebula photographs well, but is not nearly as colorful as either M-27 or M-57. As the nebula expands, it cools off, and each gas finally cools below its ionization point and stops glowing. In the case of the Helix Nebula, all that remains is hydrogen, just over 3,000 K, so this nebula appears red; see exquisite detail in July 1996's Astronomy.

In a few more thousand years, even the hydrogen will cool into a vast H I region, become transparent, and the nebula will fade visually. But we can still study these expanding shells via their radio waves long afterward. Eventually these gases will mix with the products of supernova events in the interstellar medium, to be recycled.

Observing Star Clusters

There are two seasons to observing star clusters, depending on the type of cluster. Practically all the globular clusters lie concentrated toward the Milky Way's nucleus in Sagittarius, and hence are best seen in late spring, summer, and early fall. While there are plenty of good open clusters in summer skies, too, the brightest open clusters are going to lie in our own Orion Arm of the galaxy, and will best be seen high overhead in late fall and winter.

If it is fall when you get to this portion of the text, you can still see some globulars in the west, and nice open clusters overhead as well. In Hercules, look for M-13 as a nice blur easily seen with binoculars. In dark skies with telescope 8" and up, it is well resolved into a swarm of fireflies. In really large scopes, expect a gasp from many students when they first see a well resolved rich globular. Another almost as good is M-22, just to the northeast of the top star in the teapot of Sagittarius. If you are as far south as Florida, it may be even better than the "Great Cluster of Hercules". Moderately high magnification (100-200x) is suggested to increase the sky contrast and allow better resolution of these globulars. While you are in the neighborhood, check out the fine pair of open clusters just above the tail of Scorpius, M-6 and M-7; both are faintly visible with the naked eye, and nicely resolved in binoculars. Just below Antares, M-4 is a fairly loose globular, nicely resolved in an 8".

In the summer Milky Way overhead, M-11 in Scutum is a very rich open cluster (you can see the globular vs. open distinction becomes less clear cut, with experience) that is well seen at 50x; its shape gives it the title, the "Wild Duck". A nice large binocular cluster (my wife Merry's favorite) is the "Coathanger", just south of beta Cygni in your binoculars. It sits right in that Great Rift, a long dark lane of dust in the plane of the Milky Way in Cygnus and Aquila, so the dark background makes this cluster really seem to hang in front of the nebula, in 3D. See "Meet the Summer Milky Way," in May 1996's Astronomy on page 72 for more observing highlights.

For fall, good globulars include M-15, just off the nose of Pegasus, and M-2 and M-72 of Aquarius. None are as good as M-13 or M-22, but if those have already set, they will have to do, for globulars become harder to find as we see Sagittarius setting in late fall.

But good open clusters abound all along the plane of the galaxy's disk. The Double Clusters of Perseus are easily found just east of the W of Cassiopeia with the naked eye, and look great with both binoculars and small scopes at up to 40x. These two clusters are not related; they are several thousand light years apart, and just happen to both lie in a common line of sight as seen from our solar system. They in fact are in a different arm of the Galaxy, the Perseus arm out toward the edge of the Milky Way, opposite the galactic center.

The best of the bunch is the Pleiades, spectacular in even binoculars. Keep the power low enough (no more than 20x) to keep the traditional seven sisters all visible at once. With a 4" scope or larger, look for the reflection nebula discussed in chapter 15. These are very hot, young stars; the blue color is notable in binoculars, and certainly stands out in color photos, even in star trails. Look at the contrast between the bluish Pleiades and the much older and yellower V of stars what makes up the face of Taurus east of them, the Hyades.

Many nice open clusters abound in clear winter skies. M-36, M-37, and M-38 are all nice clusters in Auriga, visible in binoculars but with individual quirks notable in telescopes at about 50x. M-35 at the foot of Gemini, just above Orion's head, stands out well in binoculars. In larger scopes at 50x, you may notice a far more distant cluster, in another arm of the galaxy, lying behind M-35; again, see if you get a sense of depth perception.

In the southern Winter sky, M-41 is a cinch to find, just 4 degrees due south of Sirius. M-46 lies east of Sirius and even contains a faint planetary nebula visible in 4" and larger scopes at 50x. East of Gemini, in Cancer, the Beehive Cluster, M-44, is faintly visible with the naked eye and almost as good as the Pleiades in binoculars.

As the Milky Way tilts south in spring, the open clusters follow it south of our horizon for the most part. Near the Southern Cross, kappa Crucis is aptly called the jewel box for the colors of its giant stars. The eta Carinae nebula is associated with a rich cluster, and many others can easily be noted by sweeping with binoculars; be sure to pack them on your next trip south.

The finest of the globulars also lies well south of the equator. Omega Centauri is the closest of these star balls, and visible to the naked eye as a fourth magnitude blur just above the southern horizon for observers south of latitude 35 degrees north. It is best seen on May and June evenings under very clear, dark skies. A 4" reflector will begin to resolve the outer portions into stars, and it is fantastic with scopes 10" or larger in aperture.

By spring the globulars start reappearing farther north as well. M-3 in Bootes is the first to show up well; look just east of Arcturus with binoculars for a small, compact blur. It does resolve well at higher power, and soon afterward, Hercules rises to take our cluster tour complete.

ANSWERS TO STUDY EXERCISES

1. The gas in the sun's interior is extremely hot, and the collisions of these rapidly moving atoms exert a radiation pressure outward to balance the gravitational pressure inward. But without the constant addition of new energy in the core by fusion reactions, the gas would eventually cool off and contraction resume.

2. When the protostar was cooler but still heating and contracting, it was far larger than the present day sun. At the cooler temperature, however, most of its energy was given off as infrared, not visible light.

3. The Pleiades has far more stars at the top left end of the main sequence, and no stars evolved off to the right in the giant stage. By contrast, the older Hyades have far few hot blue stars at the top end of the main sequence, and the brightest stars in this yellowish appearing cluster are giants already evolved to the upper right, away from the main sequence. In terms of the clusters' turn off points, the Hyades turn-off point in farther down the main sequence, so it is older.

4. The heavier the nucleus being fused, the higher the temperature and pressure needed to sustain the fusion reaction. Only the most massive stars, with the greatest gravity and capable of generating the highest temperatures, can go all the way to the formation of iron, for instance.

5. When unstable red giants shed their outer layers, the exposed core is revealed as a white dwarf star. The material shed outward may glow for thousands of years as a planetary nebula, but as the central star cools and contracts into a white dwarf, the white dwarf is left alone in space.

6. More massive stars can generate hotter core temperatures, thus form heavier elements in their final stages than can the sun. Also, as the greater gravity creates a larger core, the hotter stars are far more luminous than the sun, even while both are still just fusing hydrogen into helium as main sequence stars.

7. Globular clusters metal-poor Population II stars; open or galactic clusters contain metal-rich Popular I stars. The stars in a globular have a much narrower range of mass (typically a little less than one solar mass) than those found in open clusters. The basic evolutionary scheme is the same for Population I and II stars. However, stars with low metal abundances, such as Population II stars, form a helium core burning sequence represented by the horizontal branch on an HR diagram.

8. The sun will evolve into a moderate-sized red giant when hydrogen burning ceases in the core, causing it to heat up by gravitational contraction. The core is now rich in helium, and under great pressure it becomes degenerate. At 100 million K, helium burning via the triple alpha process begins. Turning on this "afterburner" generates much extra energy so that the excess radiation pressure may suddenly blow off the outer layers, forming a planetary nebula.

169

9. All stars will start out by burning hydrogen as main sequence stars; their will be two groups, one blue in appearance and clustered in type B at about 20,000 K and around 125 times as luminous as the sun. These heavier stars evolve faster so at first only they are visible. Later the one solar mass members turn on as G type stars, yellow in color and similar in luminosity to the sun. After about 500 million years, the more massive stars begin exploding as supernovas; in a billion years the entire upper portion of the diagram is blank. Little changes until about ten billion years, when the remaining yellow stars evolve off the main sequence, becoming red giants, then collapsing into white dwarfs.

10. The triple alpha process turns three helium 4 nuclei into a carbon 12 atom at 100 million K.

11. The different stars plotted on a HR diagram have different ages and are at different stages of their evolution. In this sense time is implicit in an HR diagram just as it is for a group of diverse people. The time, in terms of ages, becomes explicit only from computer models of the evolutionary tracks of stars of different masses and chemical compositions. These are plotted on the HR diagram and matched up with the observed HR diagrams of clusters.

12. Stars are nuclear reactors that shine by consuming themselves; once the supply of a lighter nuclear fuel is depleted, it must either turn to fusing the heavier ash it has produced in its core or, without additional radiation pressure, again gravitationally collapse.

13. In more massive stars, the opacity is less so radiative transport is more important in getting to photos to the star's surface than is the convection in our sun.

ANSWERS TO PROBLEMS & ACTIVITIES

1. A five solar mass star is about 125 X as luminous as our sun, so it is giving off about $125 \times 3.87 \times 10^{26}$ J/s = 4.8×10^{28} watts; as $E = mc^2$, then $m = 4.8 \times 10^{28} / 9 \times 10^{16} = 5.4 \times 10^{11}$ kg/ second.

2. As with Problem 1 in chapter 15, we get $L = 4 \times (3.14) \times R^2 \times$ Flux, so the flux must be $5.67 \times 10^{-8} (1000)^4 = 5.7 \times 10^4$ watts per square meter. Thus to find the stars radius, we get $R^2 = (1000) (3.87 \times 10^{26}) / (12.6) (5.7 \times 10^4) = 5.39 \times 10^{23}$, so $R = 7.34 \times 10^{11}$ km or about 4,300 AU.

3. At three billion years, the temperature should be about .987 (5780 K) = 5,700 K. At 6.6 billion years, the temperature has changed little, to just 5,860 K. At 7.7 billion years, the temperature is almost identical to the present 5,870 K. At 9.8 billion years, again the two factors practically cancel out, leaving the temperature unchanged.

4. Sin 20" = .000097, thus the size to distance ratio is 10,313 times. If it is 500 light years distant, this means the nebula is about .048 light years across.

5. Sin 30" = .000145, or a size to distance ratio of 6,900 X; for 50 arc minutes, the ratio will be 100x smaller, or just 69 X; if a typical galactic cluster is about 10 light years across, then the closer ones are about 700 light years distant, and the most distant about 70,000 light years distant.

CHAPTER 17

Stardeath

CHAPTER OUTLINE

Central Question: How do stars die, and what corpses do they leave behind?

17.1   White and Brown Dwarf Stars: Common Corpses
     A. Physics of dense gases
     B. White Dwarfs in theory
     C. Observations of white dwarfs
     D. Brown Dwarfs

17.2   Neutron Stars: Compact Corpses
     A. Degenerate neutron gas
     B. Physical Properties

17.3   Novas: Mild Stellar Explosions
     A. Ordinary Novas
     B. A Nova Model

17.4   Supernovas: Cataclysmic Explosions
     A. Classifying supernovas
     B. The origin of supernovas
     C. Supernova 1987A
     D. Supernova remnants
     E. The Crab Nebula: a supernova remnant

17.5   The Manufacture of Heavy Elements
     A. Nucleosynthesis in stars
     B. Nucleosynthesis in a supernova
     C. Other sites of nucleosynthesis

17.6   Pulsars: Neutron Stars in Rotation
     A. Observed characteristics
     B. Clock mechanism
     C. Pulsars and supernovas
     D. A lighthouse model for pulsars
     E. Binary radio pulsars
     F. Very fast pulsars!
     G. Pulsars with planets

17.7   Black Holes: The Ultimate Corpses
       A. The Schwartzschild radius
       B. The singularity
       C. Journey into a Black Hole

17.8   Observing Black Holes
       A. Binary X-ray sources
       B. Is Cygnus X-1 a black hole?

17.9   High Energy Astrophysics
       A. SS-433: a puzzle solved
       B. X-ray and gamma ray burstars
       C. Cosmic rays

Enrichment Focus 17.1: Thermal and Nonthermal (Synchrotron) Emission

CHAPTER OVERVIEW

Teacher's Notes

Planetary nebulas are some of the most beautiful astronomical objects; it would be good to illustrate discussions of white dwarf stars with photos from Hale or better yet, the newest Hubble data (see the supplement). Often the incipient white dwarf can be seen in the center of the nebula, appearing a bluish-white color due to its very high temperature.

Sirius B, the white dwarf companion of the brightest star in the sky, can be seen under good conditions with telescopes of 12" and larger. If a regular observing session is part of your course, spend some time trying to resolve it; it is a real challenge. The white dwarf is itself unimpressive, but perhaps your students will realize just how much fainter than normal main sequence stars these tiny fellows are. Both stars are about the same temperature, yet Sirius B is 10,000X fainter; this means the white dwarf must be small indeed, with a surface area 10,000X smaller and thus a radius 100X smaller than Sirius A, or about as big as our earth.

Sirius A and B pose a problem for discussion. Sirius A is a spectral type A1 main sequence star having about 2.2 solar masses. Sirius B is a white dwarf of about 1.1 solar mass. From our picture of stellar evolution, we would expect the more massive star to evolve faster, yet in the Sirius system, the less massive star is more evolved. How can this be? One possible explanation is that B was once more massive, but shed much mass in its later stages of evolution.

This comparison leads us to an important point. Students may assume that there are many black holes in the galaxy, since many stars with masses of over three suns have formed and fully evolved in the lifetime of the milky Way. However, the limiting masses for white dwarfs and neutron stars refer to the remaining mass at the end of the star's life, after the star has already shed the majority of its mass back into space as stellar winds, planetary nebulas, or supernova remnants. It is thus unlikely that many collapsed cores reach three solar masses. Black holes are probably rather rare in the Galaxy, but their total masses' role in the fate of the universe is still very uncertain.

When presenting neutron stars and white dwarfs, it helps to put some of the quantities in everyday terms. Densities like 10^{17} kg/m^3 may not mean much. This amounts to about 10,000 TONS of normal matter compressed into the volume of a pinhead, or crunching several battleships into a thimble! Do some quick calculations to find other meaningful examples. For instance, the surface gravity of a neutron star is so strong that it will not tolerate surface irregularities of more than a few millimeters. Neutron stars must be very smooth indeed.

Note that the section on high energy astrophysics incorporates different aspects of the cosmos united by their high energy phenomena (X-rays, gamma rays, and cosmic rays) related to stars and stellar evolution. Stress that the gamma ray burstars are very controversial, with no presently commonly agreed upon source.

17.1 White and Brown Dwarfs: Common Corpses

If you were to throw Jupiter into the sun, our star would become hotter, larger, and more luminous as a now more massive main sequence star. But adding mass to an already collapsed white dwarf, which has used up its nuclear fuel, just increases its gravity and causes its to shrink instead.

Back in 1984 Chandrasakhar won the Nobel Prize in Physics (rare for any astronomer) for showing that there is a limit of 1.4 solar masses that can be added and still maintain electron degeneracy. Above that limit, more mass will result in a violent collapse, then repulsive, very violent Type Ia supernova explosion.

While white dwarfs are the products of a long chain of fusion reactions, brown dwarfs never have a chance to even fuse hydrogen into helium. Refer to "Needles in a Cosmic Haystack," in September 1995 issue of Astronomy on page 50 for an update on searches for them.

17.2 Neutron Stars: Compact Corpses

A white dwarf in electron degeneracy still had atomic nuclei of protons and neutrons, immersed in a dense sea of free electrons. But the inverse beta decay process in neutron stars allows for even greater density in two ways. First the fusing of protons and electrons reduces the total number of particles present. Second, the canceling of charges ends the repulsive forces that keep particles from packing more tightly. We thus end up with a stellar core of 1.4 to 3 solar masses, compressed to the size of a city, yet with the density of an atomic nucleus, about ten million times greater than the immense density of even a white dwarf.

17.3 Novas: Mild Stellar Explosions

In contrasting novas and supernovas, it is important that the nova does not represent the end of either star in a close binary system. In fact, given time to settle down again, the system will probably create nova explosions again and again, for as long as the giant continues dumping excess hydrogen onto the white dwarf's accretion disk. T Corona Borealis, for instance, has gone "nova" three times in this century. Of course, the more violent the nova outburst, the longer it will probably take for the system to settle down and repeat the process.

By contrast, supernova explosions are a stellar "point of no return." As will be seen, their products bear little resemblance to the highly evolved stars that blew up in these more violent events. The energies of supernova are thousands to millions of times greater than even the most violent novas, and far more mass is blown back out into the interstellar medium in supernova events.

17.4 Supernovas: Cataclysmic Explosions

Type II supernovas explode because their cores form iron; no more energy can be released by fusion of heavier elements so without radiation pressure to hole their outer layers up, the cores must collapse suddenly. The energetic rebound may blow 90% of the star's mass back out into space in the expanding supernova remnant, with the remaining 10% continuing collapse into a neutron star.

Type I supernovae were probably novae several times before their final explosion. There is no such thing as a recurrent supernova! If each event blew most of the material off into space, but left the white dwarf even slightly heavier than before, ultimately, the white dwarf might accrete enough mass to pass the Chandrasakhar limit of 1.4 suns. But, instead of being able to collapse down to a stable neutron star as did the Type II supernovae cores, the Type I supernova apparently completely destroys the white dwarf, blowing all of its material into space. Since there is no power source inside, Type I supernovae remnants fade quickly.

The presence of neutrinos from Supernova 1987A was expected, and made us realize the neutrinos must be essentially massless. If they had any mass, they could not have traveled at the speed of light and arrived before the light of the supernova did. Did they, in fact, travel faster than the speed of light? No; neutrinos can cut through the outer layers of the collapsing star easily, but the visible light comes later when the shock wave of core collapse reverberates all the way up to the surface. In fact, the time delay is important in helping us understand why this supernova did not get as bright as we expected with Type II supernovas. In the traditional model, the star which explodes is either a bloated red supergiant, much like Betelgeuse, or like the supernova discovered in M-81, (discussed in Sky & Telescope, December 1993, page 30). That supernova was a K0 supergiant. The article does a good job of distinguishing between Type I and II supernovae, as well. When the shock wave reaches its huge photosphere, the surface area is immense and the supernova may temporarily outshine all the rest of its galaxy. But, this would mean that several days, or even a week, would be needed for the shock wave to reach the red giant's surface. The shorter time delay is exactly in keeping with the fact that before the star went supernova, the star was a B3 supergiant, much more like Rigel than Betelgeuse. Such a hot supergiant would need to be only about 20-30x larger than the sun--not hundreds of times larger, as it would for a star like Betelgeuse. But, since it was smaller, the photosphere did not brighten as much as we had hoped, and Supernova 1987A did disappoint many by never getting brighter than Polaris. Our first estimates had predicted that it would get almost 100 times brighter--at least as bright as Jupiter, if not Venus.

Remind students that Type II remnants remain brighter far longer for they have pulsars in their centers, providing synchrotron radiation to keep the nebular gases excited. As the entire white dwarf blew up in Type I supernovas, their unpowered remnants fade quickly into ghostly shells like Figure 17.22, and are best observed via their radio transmissions. Interestingly, its was recently found that the majority of cosmic rays reaching earth come from the remains of SN 1006, a type Ia SN.

17.5 The Manufacture of Heavy Elements

Our models of the Type II supernova cores can go a long way to explaining the abundances of the heavier elements. Iron and nickel are the most common really heavy elements for the can be produced as the end products of both fusion of lighter elements and radioactive decay of still heavier elements. What the star has plenty of time to produce, it makes in cosmic abundance, such as carbon, oxygen, calcium, silicon, sulfur, and iron.

174

But what must be made in the fraction of a second, such as all the rare elements beyond zinc, are going to be very rare--there is not time to make much of them! To confirm this, merely look at the periodic table; once you pass zinc, all elements that follow are more than a thousand times less abundant that iron. Thus supernovas explain why iron costs pennies a pounds, and gold $400 an ounce. It is also the tremendous violence of that fraction of a second core collapse that breaks apart the atomic nuclei, spilling neutrons everywhere to create a host of radioactive isotopes in the r and s processes. As these radioactive nuclei, such as Al 26, decay many years later, they impart some of the stored energy trapped long ago in their supernova formation to the core of the planet which they accreted into. In a sense, the eruption of a volcano is just an echo of the far greater violence of the supernova which forged the earth's Uranium 238 billions of years ago.

While the common, everyday elements are not formed in the supernova's core collapse, we are still its children. What good would the critical elements like oxygen, carbon, et c. have done us had they remained inside the red giant's core? The supernova stage turns the evolved giant inside out, blowing back into the interstellar medium the elements which would become us. We are children of stars that had to die so that we might live...

17.6   Pulsars: Neutron Stars in Rotation

How could one hope to see a star no bigger than a city? Yet we have seen a few, and heard a lot more! As the core shrinks, the conservation of angular momentum causes the neutron star to spin faster and faster. Sagan's ice skater analogy in Cosmos 109 is very good here. Also remember the last common thing formed before the core collapse was iron. This outer "crust" of iron, plus a spin period of hundredths of a second, generates a dynamo of often hundreds million times the earth's magnetic field strength.

As Uranus and Neptune show us, there is no cosmic law that magnetic fields must align themselves with the body's rotation axis. As the neutron star spins, the lighthouse model calls for the poles to go flashing past us twice every rotation. This bipolar ejection pushes bursts of synchrotron radiation outward toward us like water out of a sprinkler. Zeilik notes that these are going to be very dependent on the axis's orientation; like eclipsing binary systems, we probably miss the vast majority of pulsars. Also note that each pulse carries off with it some of the angular momentum of the rapidly spinning pulsar. Thus as they age, pulsars will slow down, and their intensity fade (happen to you, too, young'uns....). If we carefully graph the rate at which the pulsar is slowing down, we can extrapolate back into time to when the pulsar was spinning very rapidly, we can estimate when the supernova occurred. We get very good agreement between the Chinese observations of 1054 and the slowing down of the Crab's pulsar, for instance.

The idea of planets forming in the supernova's aftermath around a neutron star is amazing, but several examples have now been studied. See the January 1996 issue of Astronomy on page 50.

17.7   Black Holes: The Ultimate Corpses

Note again to your students that the sun is NOT a potential black hole. The gravitational force needed to compress a core to that extreme can happen only in the most massive stars. If 90% of the star's mass were blown out in the supernova event, it could take a 30 solar mass star to leave just three solar masses in the core and form a black hole. There are not going to be but a handful of such hypergiant stars in the Galaxy at any given time, so there may be few such collapsed cores.

17.8 Observing Black Holes

We have already discussed how the doppler shifting of a visible star in chapter 14 could tell us of the presence of a massive dark body, such as Cygnus X-1 (or even lower mass planets, like those found in 1996). Remember also that a black hole is a tremendous gravity well, accelerating infalling matter from the companion star inward as speeds approaching the speed of light. In fact, the "event horizon" is that boundary at which the matter is accelerated up to c, and time freezes (hence event horizon). Collisions at such velocities create temperatures of millions of degrees. Wein's Law reminds us that means a not of X-rays are created; Cygnus X-1 is the third strongest X-ray source in the entire sky.

17.9 High Energy Astrophysics

Gemininga, the most intense gamma ray source in the sky, appears to also be the nearest pulsar, but with its pulsed beams not falling in our line of sight. Apparently the gamma ray emission is not nearly so directional as the longer wavelengths. See the December 1993 issue of Astronomy.

The almost uniform distribution of gamma ray burstars over the entire sky is discussed in both the December 1995 issue of Astronomy and the September 1996 issue of Sky and Telescope. It is possible there are many "invisible" neutron stars around us, such as Gemininga, but we become aware of them only when a comet or asteroid violently hits their surface, according to one theory.

ADDITIONAL RESOURCES

The slide 17.21 and 17.26 and transparencies 17.1, 17.4, 17.12, 17.15, 17.16, 17.23, F.13, 17.29, 17.33, 17.36, and 17.39ab add color and graphics to your class.

SUPPLEMENTARY ARTICLES

1. "Star Death," Astronomy, February 1988, page 6.
2. "The Origins of the Elements," Astronomy, August 1988, page 18.
3. "Exotic Pulsars," Astronomy, December 1988, page 22.
4. "The Aftermath of Supernova 1987," Astronomy, February 1989, page 40.
5. "Hunting for Supernovae Down Under," Astronomy, November 1989, page 94.
6. "Observing Planetary Nebulae," Astronomy, September 1991, page 76.
7. "A Burst of Gamma Rays," Astronomy, November 1991, page 46.
8. "Will Supernova 1987A Shine Again?," Astronomy, February 1992, page 30.
9. "The Best Black Hole in the Galaxy," Astronomy, March 1992, page 30.
10. "Lost and Found: Pulsar Planets," Astronomy, June 1992, page 36.
11. "Giving Birth to Supernovae," Astronomy, December 1992, page 46.
12. "Brightest Gamma Ray Burst Seen," Astronomy, August 1993, page 19.
13. "The Star That Blew a Hole in Space," Astronomy, December 1993, page 30.
14. "Ashes to Ashes," Astronomy, May 1994, page 40.
15. "Where Have All the Black Holes Gone?," Astronomy, October 1994, page 50.
16. "Inside the Crab Nebula," Astronomy, December 1994, page 42.
17. "Gamma Ray Burstars: Near or Far?," Astronomy, December 1995, page 56.
18. "Second Chance Planets--Orbiting Pulsars!," Astronomy, January 1996, page 50.
19. "Fastest X-Ray Gun in the West," Astronomy, September 1996, page 28.

ADDITIONAL DISCUSSION TOPICS

Nova Cygni 1975 and Me

On August 27, 1975, I was visiting my parents at their farm north of the small town of DeFuniak Springs in the Florida Panhandle. I decided to take advantage of the clear night to take some photos of the northern Milky Way for use in the Pensacola Junior College Planetarium. I was fairly familiar with the bright constellations. I noted, as usual, the appearance of the three bright stars of the summer triangle as they appeared overhead as twilight fell. I set up the telescope and camera. But wait, there, north of Deneb, there should be no third magnitude star.... I quickly checked the star atlas, and indeed no star that bright was plotted there. Thinking I might have an earth satellite, I waited for motion, but observed none. Could it be a planet? No, planets of the summer sky will lie far to the south, in the zodiacal constellations Scorpius, Sagittarius, or Capricorn; none of you were born under the sign of Cygnus. At high magnification, my mystery object was a starlike point of light, not a weather balloon or flaring comet. It also showed no motion relative to the background stars; this was not an object within the solar system but a vast and sudden change in a distant star. Deciding to put my name on the line, I called the IAU number for reporting comets (or any other fast-breaking news) at (617) 864-5758 and reported my discovery of a new nova; I gave its position in Cygnus and noted that as of 8:15 PM CDT, its apparent magnitude was +2.8 and rising fast. I was the first American to report the brightest nova of my lifetime. On every light curve ever published for this nova, that one data point at magnitude 2.8 is my report. I then called up several other observatories, including rousing my graduate astronomy professor, Dr. John Oliver of the University of Florida, out of bed about 10 PM EDT. In Gainesville, it was in the middle of a thunderstorm, but from my excitement he knew he had to get dressed and drive 30 miles south to the University's Rosemary Hill Observatory. As he pulled into the observatory parking lot, the clouds miraculously parted; he looked up to see the new star just north of Deneb. His spectra that morning were the first taken of a nova on the rise like this, and were vital in establishing an estimated distance of 3,000 light years to the explosion.

The next evening I went out, well before sunset, to scan the northeastern skies in hopes I had discovered a supernova, growing bright enough to spot in broad daylight. This has been observed before, with the supernovas of 1006, 1054, and 1572, but my event was not such a gigantic explosion. In fact, although it was to peak as the sixth brightest object in the summer sky the following evening, my nova would fade rapidly, dropping below naked eye visibility within two weeks of my discovery. Still, it was fun while it lasted, and this excitement is part of the reason that people become astronomers. We never know when something grand and wonderful will be unveiled before our eyes. If you or your students will take the time to learn the constellations well, they too may looking in the right place at the right time to witness such a stellar explosion. Your chances of making such a discovery are better of course with binoculars, and in general the far greater density of stars in the plane of the Milky Way means those constellations are site for many more nova and supernova discoveries; it all goes to show that it can pay to look up. Perhaps a comet discovery, which gets your name attached to the comet forever, is better, but still discovery of a nova can get you in the headlines, as it did for Spanish amateur Francisco Garcia when he discovered Supernova 1993j in the galaxy M-81. He was using 10" reflector, smaller than many college observatories use with their classes.

Historical Supernovas

While Zeilik mentions the supernova of 1054 AD that made the Crab Nebula, there have been other historical supernovas that far outshone it in spite of Sagan's account in Cosmos 109. The one which created the Gum Nebula about 7,000 years ago was nearby and quite brilliant, perhaps outshining the full moon. The Sumerians wrote of this Great Star of Ea in some of their first cuneiform tablets--it may have literally inspired writing, as people sought to record the great spectacle their ancestors had witnessed. The expanding supernova remnant from it has now expanded to about 30 degrees across, and at its center the Vela Pulsar still powers the expanding remnant. It is coming to get us; in perhaps 10,000 years our descendants will know what it is like to live inside a supernova remnant. Probably the magnetic force field of the sun will repel most of the supernova shock wave, and little damage will occur on earth, but it still could set up some interesting auroral displays on the edge of the solar system.

In 1006, a supernova in Lupus was observed from southern China and southern Europa. A monk in a Swiss abbey records it barely climbing above his southern horizon through a break in the Alps for just a few minutes each evening. Because of its low altitude and great brightness (at least as bright as the quarter moon), it put on a fine fireworks show with atmospheric distortion breaking down the supernova's light into spectral shafts of topaz, emerald, ruby, and sapphire. What a spectacle! In 1996, it was found to be the main source of cosmic rays striking the earth as well.

In 1572 Tycho saw the familiar "W" of Cassiopeia vastly distorted with a supernova growing brighter than Venus and visible for months in broad daylight. In 1604 Kepler noted another in Ophiuchus which grows brighter than Jupiter for several weeks. Yet both of these recent supernova left behind no glowing remnants like the Crab and Gum Nebulas; these were both Type I supernovas, and blew themselves up entirely, leaving behind no collapsed core objects such as the neutron stars in the middle of the Type II remnants.

DEMONSTRATIONS AND OBSERVATIONS

Observing the Crab Nebula

The only supernova remnant bright enough to be easily seen with amateur telescopes in the northern hemisphere is M-1, the Crab Nebula of Taurus. It can be seen with large binoculars or a small telescope as an oval blur about a degree NE of beta Tauri, the southern horn of the Bull. This is the same place the Chinese reported the naked eye supernova in 1054, as described on page 379. But this object looks far better on photos such as figure 17.27 than it does through the eyepiece; it is easy to see why Messier mistook it for the slightly elongated coma of a new comet, and waited hours for it to move. It was his frustration with this object that caused him to start his famed catalog of objects comet hunters should avoid wasting time with.

As Zeilik notes, the radiation from the spinning pulsar is synchrotron, and thus polarized. Try holding a Polaroid filter in front of the eyepiece; do you find the nebula appears different at some orientations than at others? The tendrils of red hydrogen, however, are visually invisible; even the famed pulsar, which can be seen with a 16" telescope, does not flicker visually. It is pulsing 30 times a seconds, and 30 Hz is a little to fast for our retina and brain to record as separate pulses. But perhaps your class will understand that just to find the one visible tombstone of a stellar suicide a millennium ago is no mean accomplishment.

178

ANSWERS TO STUDY EXERCISES

1. A white dwarf has a high surface temperature, a mass of 1.4 solar masses or less, and its about as large as the earth. It consists of a degenerate electron gas in which the normal electron orbitals have been utterly crushed. The electrons move freely among the nuclei, now in the crystalline structure with a density a million times greater than normal matter. Its luminosity arises from the slow release of gravitational energy as it shrinks; having used up its fusionable fuels, no more fusion reactions can occur.

2. A neutron star is even denser than a white dwarf, for most of the protons and electrons have been fused together to form a core of pure neutrons, packing into a degenerate gas. A neutron star is between 1.4 to 3 solar masses, yet this collapsed core is only about ten kilometers across. The surface gravity of neutron stars is billions of times greater than on earth. Just after they are formed in supernova events, neutron stars will be spinning very rapidly due to shrinking to a small size in collapse while conserving angular momentum.

3. We can telescopically observe numerous white dwarfs close to the earth, such as Sirius' companion. In these binary systems, we can directly measure the mass of both stars. From their observed high temperature and low luminosity, we can calculate their small sizes. The evidence for neutron stars is less direct, chiefly from the observed properties of pulsars, thought to be rapidly rotating neutron stars whose strong magnetic fields imply tiny sizes.

4. Almost all the elements beyond hydrogen and helium were either made or at least liberated into the interstellar medium in supernova events; only the jovian planets and stars would be there, and they would certainly not be as visible without their heavier compounds.

5. A .5 solar mass star will make a white dwarf, while a two solar mass star will collapse into a neutron star, instead. In reality, the heavier star will probably lose enough matter in its final stages that it would probably become a white dwarf as well.

6. Both photographs of the filaments and their doppler shifts indicate the nebula is rapidly expanding; extrapolating back into time, this expansion seems to have begun at the same time the Chinese observed the visible supernova, in 1054 A.D. The plotted position of the event by the south horn of Taurus agrees with the placement of the Crab in the sky. The pulsar in the center of the Crab is slowing down at a rate consistent with its formation about 950 years ago. Also the energy given off by the Crab is highly polarized and synchrotron in form, not typical of a glowing H II region.

7. A black hole censors light within its event horizon, preventing us from getting information about this densest of all collapsed stellar cores. Matter, too, is trapped within this region.

179

8. The region around the black hole usually will include a rapidly spinning accretion disk, where infalling matter will be accelerated up to close to the speed of light and thus generate temperatures of millions of degrees; at this temperature, X-rays are the main form of energy.

9. Light can't escape out of the event horizon, nor could we get any information from a probe falling into a black hole once it crosses this horizon.

10. The small time intervals between pulses imply very rapid rotation, some at hundred of rotations per second. Only a neutron star is small enough and strong enough to spin so rapidly and not fly apart. Also, as each pulse radiates away some of the rotational energy, older pulsars spin slower and become weaker as the magnetic fields fade in intensity. We can even use the rate at which they are slowing down as an indication of their age.

11. Accelerations of electrons in magnetic fields generate synchrotron radiation at a wide range of wavelengths. The electrons come from the surface of the neutron stars.

12. Cosmic rays from solar flares come from the direction of the sun itself, while those from other sources in the galaxy come from all directions in space, but with a greater concentration along the galactic plane.

ANSWERS TO PROBLEMS & ACTIVITIES

1. Escape velocity = $(2 \, GM/R)^5 = (2 \times 6.67 \times 10^{-11} \times 2 \times 10^{30} / 10^4)^5 = 1.6 \times 10^8$ m/sec, or about half the speed of light.

2. For a spherical blackbody, $L = 4 \times (3.14) \times R^2 \times F$, where the flux, $F = 5.67 \times 10\text{-}8 \times (T)^4$;
$F = 5.67 \times 10^{-8} \times (1500)^4 = 2.9 \times 10^5$ Watts per square meter. Thus the luminosity L is given by L $= (12.6) (7 \times 10^7)^2 \times (2.9 \times 10^5) = 1.8 \times 10^{22}$ Watts, less than .0001 of the sun's.

3. Sin 6.5' = .00189, so the Cassiopeia remnant is 529 times more distant than its diameter. Thus at about 9,000 ly, its diameter must be about 9,000/529 = 17 light years.

4. If 2000 km/sec corresponds to .2 arc seconds per year, then in a year, the nebula expanded by 6.3 $\times 10^{10}$ km. Finding the size to distance ratio, sin .2" = .000001, so the nebula is 1,031,324 times farther away, or 6.5×10^{16} kilometers, or about 6,900 light years distant.

5. The sun's present density is 1,400 kg/m^3, and its present radius is 7×10^6 km; as the volume is proportional to R^3, if the radius is reduced to just 10 km, then the density is increased by $(7 \times 10^7)^3$ $= 3.43 \times 10^{23}$ times greater, or about 4.8×10^{26} kilograms per cubic meter.

6. At the event horizon of a black hole, the escape velocity = $c = 3 \times 10^8$ m/s; thus putting your mass (about 100 kg) into the escape velocity formula, your Schwartzchild radius = $2GM/c^2 = 2 \times 6.67$ $\times 10^{-11} \times 100 /(3 \times 10^8)^2 = 1.5 \times 10^{-25}$ meters.

CHAPTER 18

The Evolution of the Galaxy

CHAPTER OUTLINE
Central Question: What evolutionary processes induce the structure of the Galaxy?

CHAPTER OVERVIEW

Teacher's Notes

This chapter may well be the most difficult and most confusing in content so far. Non-science majors will probably not have had any exposure to wave motions in general, so it may be unreasonable to expect them to apply an abstract concept such as density waves to galactic structure. This chapter does apply ideas presented and repeated throughout the text, particularly Kepler's Third Law and the Doppler effect. Review these concepts (again) if necessary, and use them to establish some basis for familiarity with the new material.

Viewing photographs or slides of other spiral galaxies should help students to picture the Galaxy and to accept that the spiral structure is a reasonable or possible interpretation of the data. The next step might be to convince them that the spiral can't be material arms because they would wind up around the nucleus within the time of a few rotations of the sun around the Galaxy. A demonstration might help here; one which can be done, but which may be a bit messy to set up, involves spinning or turning a cork floating in the center of a shallow pan or petri dish containing a viscous liquid, such as glycerine. Lines radiating outward from the cork may be drawn on the liquid using food coloring. As the cork is turned, the straight lines twist and wrap in a spiral around the cork. If the pan is transparent, the whole affair can be placed on an overhead projector. This setup could be used at the same time to illustrate pitch angle.

Use everyday examples, as in the text, to illustrate density waves or compression waves. An 8-mm film loop entitled "Non-Recurrent Wave Fronts" shows a compression wave propagating through crowds of people. The wave front is clearly visible and sweeps quickly through the crowd.

18.1   The Galaxy's Overall Structure

When Galileo turned his early telescope toward the Milky Way in 1611, he resolved it into myriads of stars. Since then better and better telescopes and other detectors have given us some insight into the great system of stars, gas, and dust that our solar system is a part of. But it is often not easy to see the forest for the trees; Herschel's star counts placed us near the middle of a disk only 10,000 light years across. Even today, there is no consensus on exactly where in the spiral arm structure our solar system lies or even what to call the spiral arm closest to us; a plastic model of the Galaxy I use in class puts us in the "Carina-Cygnus" arm, while a slide diagram from an old text takes the safest approach, placing us in the "Local Arm."   Several books now say we lie in either the "Orion Arm", or a smaller "Orion Spur," referring to the part of the arm we have already passed through. In his planetarium show, "Galaxies," Timothy Ferris places us on the inside edge of the more distant Perseus Arm. I personally prefer the "Cygnus Arm," since I like to look ahead of me, where we are going, instead of back where we have already been.

Beyond that, there is general agreement that the Milky Way is a spiral galaxy, but disagreement over whether or not it is a barred spiral galaxy. The nucleus of the galaxy is being probed with new instrumentation, which find hints that our nucleus was once a quasar whose eruptions may have produced jets which in turn led to the formation of our satellite galaxies, the Large and Small Magellanic Clouds. In the summer of 1995 came the announcement that a third smaller satellite, twice as close as either Magellanic Cloud, lies in the direction of Sagittarius. Even close to home, there is still so much to learn about the true nature of our home Galaxy.

There is no consensus as to what to call our present location in the galaxy. Some books say we lie in the Local Arm, others in the Cygnus-Carina Arm, others the Orion Arm (or Spur), others the Cygnus Arm, etc. Zeilik puts us in the Orion Arm, between the outer Perseus Arm of the northern winter sky and the Sagittarius Arm of the southern summer sky. As we are looking toward downtown in summer, this explains why the summer Milky Way is so rich while the opposite direction and season have us looking out to the edge, with not so much bright material in that direction. Even the exact distance from the center to the sun is not so certain; this text gives it as 30,000 light years, but others say it may be closer to 25,000 light years.

18.2 Galactic Rotation: Matter in Motion

Why did the initial models of the galaxy put us near the center? Herschel's model was based on real data; he tediously made star counts at high power in many directions in space. He found the number of stars dropped greatly as he moved above or below the plane of the galaxy, but that we found about equal numbers of faint stars in all directions along the galactic disk. This naturally led him to the conclusion that we lay near the center of a disk only 10,000 light years across. His problem was that he did not understand the role of the dark nebulas; he ran across many of them, and attributed the lack of stars in their directions as being holes in the heavens, due to a real lack of stars. Only after Barnard's long exposures revealed the dust lanes silhouetted in front of the star clouds did astronomers begin to realize that such dark clouds were blocking our visual view of over 90 % of our home galaxy.

Shapley succeeded because he avoided the dusty disk. The globular clusters lie in a spherical array around the galactic nucleus, with many well above the plane and obscuring dust lanes. Thus he could look clearly at globulars lying over 70,000 light years distant, on the other side of the galaxy. As Figure 18.7 and our discussion of clusters in chapter 16 show, the globulars are almost all found in the sky's summer hemisphere, centered toward Sagittarius.

Note the mass discrepancy; about 100 billion solar masses lie between us and the center, yet the rotation curve implies a total mass of over a trillion suns. Again, our eyes reveal only the tip of the galactic iceberg.

18.3 Galactic Structure

In trying to explain the period-luminosity relation, remind the students that these are unstable, giant stars who are switching from one fusion cycle to another. The more massive (and luminous) the cepheid variable, the longer it takes for it to get its act together. If we know how long it takes for the cepheids to pulsate, we can estimate their luminosity and absolute magnitude, and then use the distance modulus formula to estimate the distances to the cepheids and the clusters (or even external galaxies) to which they belong.

In trying to map the galaxy, again note how miserably our eyes fail us. We visually see only about 10 % of the Milky Way. This is why the radio mapping of H I regions at 21 cm is so important; we are trying to find out how many spiral arms the galaxy has, and if the Milky Way is the normal or (as the latest evidence suggests) barred spiral. If the Milky Way is in fact a barred spiral, then the orbit of the solar system may be considerably more elliptical than previously thought, as the solar system might come considerably closer to the galactic nucleus than its present position. The collisions with the smaller companion in Sagittarius may help account for the Milky Way's structure.

18.4   Exploring Galactic Structure by Radio

If students fail to see how longer waves can penetrate the dust clouds, point out that a dark rain cloud or dust storm does not interfere with radio or television reception.

Note how little a shift in the spectral line of H I at 21 cm occurs due to galactic rotation. The waves are shifted only about .1 and .2 cm, according to Figure 18.13. But this gives a doppler shifts of .1/21 to .2/21, or between .5 and 1% of the speed of light. These arms are traveling, relative to us, at between 1,400 and 3,000 km/sec.

The mapping is hindered by any local disruptions of the overall rotation of the arms. Hence a supernova shock wave may disrupt the smooth flow of gas and make the speeds and distances to that part of the galaxy suspect. This is a prime reason that different observers come up with different details for the arm structure. It is also possible that collisions and interactions with other galaxies have disrupted the disk somewhat, and that some of the motions observed are not in the disk, but either above or below it. The Sun, for instance, lies several hundred light years below the exact galactic plane, as shown by our viewing the southern portion of the galactic bulge better than the northern, which is blocked by the dust clouds in the disk. It is the "southern bulge" that visually marks the Sagittarius companion galaxy, now being tidally devoured by our giant spiral.

It is worth noting that the giant molecular clouds are observable only in microwave and radio wavelengths. If you had a nebula so hot that it glowed brightly, like an H II region, it would be so hot that the organic type molecules would be broken apart and not detected. You need a cold and dark place to get the atoms to collide often enough and slowly enough to stick together and make the molecules we detect in these long wavelengths. Since the molecules are themselves cold, perhaps only a few degrees K, then any radiation they emit must be of low energy, or long wavelength accordingly. These giant molecular clouds are fairly compact sources, so they are easier to pinpoint on our galactic maps than the broad distribution of hydrogen in the H I regions along the spiral arms.

18.5   The Evolution of Spiral Structure

In getting students to transfer the spiral density wave model to observations, note that if the traffic is backed up, the density of interstellar material is great enough to produce more massive stars faster than normal, hence leading to the birth of the stellar associations whose bright blue stars light up the leading edges of the spiral arms. The supernova shock waves these stars produce when they die also cause the interstellar medium to be compacted, leading in turn to formation of less massive stars like our sun as well.

18.6   The Center of the Galaxy

The Walt Disney movie, "The Black Hole", made this gravitational well appear most sinister. So it might be up close, but the black hole in our core has the far more appealing role of providing some of the gravitational glue that holds the galaxy together. It seems likely that such gravitational collapses are critical to the formation of practically every sizable galaxy.

If the galactic nucleus is hidden from us visually, why is it so detailed in infrared maps? It would seem that less energetic infrared photons would have even more trouble reaching us. But remember that when a photon of light is absorbed by a dust grain, the grain in turn gains kinetic energy--it gets hotter. It then radiates this energy away as heat, or infrared radiation that can pass through the dusty disk and reveal the pancake shape of the Milky Way's disk.

184

18.7 The Halo of the Galaxy

The two most likely candidates presently for the dark matter are at opposite ends of the stellar mass range, brown dwarfs and black holes. Brown dwarfs are of course not massive enough to shine, but may be very abundant; only future infrared surveys will tell. Black holes are collapsed cores, too massive to shine as pulsars. They too may be very abundant, but unless we are lucky enough to see one interacting with a close companion, practically undetectable.

18.8 A History of Our Galaxy

In considering the evolution of the galaxy, I think to compare it with the evolution of the solar system. The initial condensation was in a spherical array, forming the halo stars and globular clusters around the edge of the galaxy, the Oort Cloud and the comets around the outskirts of the solar system. As condensation continued, the rotation of the material created a disk; on the grand scale, the galactic plane, closer to home, the ecliptic. In that disk, the more massive objects condensed, to become the associations of giant stars that outline the galactic plane today, and the bright planets that mark the ecliptic in our back yard.

ADDITIONAL RESOURCES

Transparencies 18.1, 18.4, 18.5, 18.7ab, 18.9, 18.12, 18.13, 18.15, and 18.23 will reinforce your students' attempts to find themselves in the Milky Way.

Several scientific companies sell a plastic model Milky Way about 15" across. It shows the disk, nuclear bulge, stellar populations, and our position well, but the naming of the arms is not in agreement with this text. To illustrate the newly found companion, add a 3" long whisp of cotton batting, on the edge opposite the solar system. Pass it around among your students during class.

SUPPLEMENTARY ARTICLES

1. "A Bigger and Better Milky Way," Astronomy, January 1984, page 6.
2. "Globular Clusters, Ancients of the Universe," Astronomy, May 1985, page 6.
3. "The Stellar Populations," Astronomy, October 1986, page 106.
4. "Is the Milky Way an Interacting Galaxy?," Astronomy, January 1988, page 26.
5. "Unveiling the Hidden Milky Way," Astronomy, November 1989, page 32.
6. "The Magnetic Milky Way," Astronomy, June 1990, page 32.
7. "Strange Doings in the Milky Way's Core," Astronomy, October 1990, page 39.
8. "Life Near the Center of the Galaxy," Astronomy, April 1991, page 46.
9. "Journey to the Center of the Galaxy," Astronomy, July 1991, page 74.
10. "Galactic Archeology," Astronomy, July 1992, page 28.
11. "Journey into the Galaxy," Astronomy, January 1993, page 32.
12. "In the Beginning: The Globular Clusters," Astronomy, October 1993, page 40.
13. "Intruder Galaxies in our Neighborhood," Astronomy, November 1993, page 28.
14. "What lies in the Center of the Galaxy," Astronomy, April 1995, page 32.
15. "Globulars and the Age of the Universe," Astronomy, February 1996, page 44.
16. "Meet the Summer Milky Way," Astronomy, May 1996, page 72.
17. "Barnard's Magnificent Milky Way," Astronomy, June 1996, page 32.

ADDITIONAL DISCUSSION TOPICS

The Galactic Year and the Ice Ages

There is not any consensus as to the length of the Galactic Year, the time for the solar system to revolve around the Galactic Nucleus. It is estimated at between 200 and 300 million earth years, depending on how fast you think we are moving and how far out from the center we lie, both figures being debated at present. Zeilik mentions we have been around the Milky Way about 20 times since the solar system's formation, so he chooses an average of 250 million years.

As we revolve around the center of the galaxy, the solar system will pass through the dusty spiral arms; we are not part of them nor fixed in relation to them. The arm structure of the galaxy in our vicinity would have looked very different when the dinosaurs were still on the earth. In fact, some think that spiral structure is a sometime thing, with the arms brought to prominence by the tidal perturbation of other passing galaxies, then lapsing back to almost elliptical appearance over time.

If these arms are normally as dusty as both the Orion region from which we have just emerged and the Cygnus region that we are heading toward, this interstellar dust may have climatic implications. This dust could attenuate the sun's light, chilling the planets into ice ages. Even if there wasn't enough dust to directly block the sun's light, dust pulled inward by the sun's gravity might chill the sun's photosphere, again leading to ice ages.

We are currently in such a cold period and have been for at least the last two million years. This is not normal during most of the planet's history, for paleontologists think that the entire globe has been free of standing glacial ice, even in the polar regions and on mountaintops, during most of the planet's history. Such cold periods seem to recur in intervals of about 250 million years, suspiciously near the length of the Galactic Year.

Where is Most of the Universe?

The fact that the galactic rotation curve differs from keplerian motion is testimony that unlike the solar system, where most mass lies in the central sun, the majority of matter in our galaxy lies well beyond the orbit of the solar system. Yet our eyes tell us the brightest part of the galaxy lies toward the center. But again our eyes deceive us. As we look up, we see a sky dominated by bright blue stars. If we look at other similar spirals, we find most of their light comes also from such bright blue stars of high mass and short life spans.

Yet in our census of stars around the sun, we found our star the fourth most luminous of the forty closest suns. When we sum up the total masses, the little K and M main sequence stars make up the majority of the mass of stars close to us just by shear weight of numbers. Does this trend continue? What if for every red dwarf star we do see, there are dozens of brown dwarfs too low in mass to shine, but still adding more mass to the galaxy? It is a paradox; our eyes see a galaxy dominated by high mass stars, but the rotation curve reveals a galaxy whose motion is dominated by objects even less massive than the sun.

The November 1993 issue of <u>Sky & Telescope</u> has a interesting news note on the search for MACHOs (Massive Compact Halo Objects) on page 9. Apparently gravity lensing by brown dwarfs has been detected in at least three cases, suggesting these may be fairly common, perhaps amounting to the majority of the missing mass. I suspect that both high and low mass bodies will end up playing major roles in solving this mystery.

186

Radio Astronomy and World War II

The radio emissions from the galactic center (Sagittarius A to radio astronomers) are the strongest radio source in the sky. Karl Jansky of Bell Labs discovered them by accident in the 1920's and Grote Reber mapped them in some detail during the 1930's. During World War II, German bombers would fly in toward London from the direction of Sagittarius rising, using the radio static from the galactic core to hide from British radar. After the war, much progress was made by using German and allied surplus radar antennas as radio telescopes; the Dutch in particular did a lot of early research because their countryside was littered with the German radar network used by the Luftwaffe. Many of the radar technicians on both sides were to turn their expertise to peaceful uses and become the first generation of radio astronomers.

The Milky Way as a Quasar

The turbulence of the radio observations in figure 18.18 suggests that even today the core of our galaxy is not a placid place. It may have been far more violent in the past. If the black hole at our galaxy's center was better fed, it would have been far more energetic. Many think that quasars are in fact such well-fed black holes, typical of the earlier stages of galactic evolution. Such violent nuclei in turn often spew out great jets of material, such as the bipolar flows we met in both the birth and death of stars but now on a galactic scale. Many astronomers now consider that the Magellanic Clouds, far younger than the Milky Way itself, are the product of material thrown out in a violent quasar stage perhaps around six billion years ago. Could the material from the Sagittarius companion, fed inward toward the nucleus, even restart the "quasar machine?"

They are still connected to mother Milky Way by the Magellanic Stream, an umbilical cord of hydrogen from the nucleus of the Milky Way passing through both of the Clouds.

This in turn may have implications to life elsewhere in the Galaxy, and in the universe as a whole. In its quasar stage, the nucleus may become such an intense source of X-rays and gamma rays that it would sterilize the entire galaxy. The intense radiation would destroy all organic molecules and life forms exposed on the surface of any planet. Perhaps ET has not come calling because our solar system was one of the first planetary systems to form after the Milky Way calmed down enough to give organic chemistry a chance.

Population III?

Because even the oldest known population II stars contain at least traces of heavier elements than helium, some astronomers argue that there was an initial population III, giant stars made of nothing but the pristine hydrogen and helium made in the Big Bang. Such stars are long gone; they would have evolved rapidly, blown up as supernovas, and showered the young galaxy with the traces of heavy elements we find in even the oldest population II stars. And since they were massive, their collapsed cores would have become black holes, the missing mass to the halo.

The idea of a short-lived, massive initial population might go a long way toward explaining the high luminosity and blue color of the very youngest galaxies we observe at the edge of the visible universe. As already mentioned on the previous page, if these massive stars did collapse into black holes, these would be very hard to detect in the sparsely populated outer regions of the halo (also called the Galactic Corona in some books), but still help explain the missing mass problem. Keck telescope observations in 1996 indicated that there may have indeed been such a population.

DEMONSTRATIONS AND OBSERVATIONS

Observing and Photographing the Milky Way

The Milky Way is faint, so the first rule is to get away from light pollution as much as possible. Since the brightest portion lies in our southern sky, perhaps the beach might be a good choice. Allow time for your eyes to become dark adapted, usually at least 20-30 minutes. Of course the night should be as dark as possible, with the moon well below the horizon. The Milky Way on spring evenings lies around our horizon, with the galactic pole at our zenith. This may be a fine season to look beyond the Milky Way to the rich Realm of the Galaxies in Virgo, but it is not good for seeing our home galaxy. In early summer the brightest section of the Milky Way rises along our eastern horizon, with the center lying in the direction of the teapot of Sagittarius, just above Scorpius' tail. In fact, the teapot's spout even seems to have a cloud of steam (the Sagittarius star cloud) pouring out of it, pointing to downtown Milky Way.

We have already noted that binoculars are superb instruments for scanning the galaxy's clusters and nebula. They will also reveal smaller dust lanes in Scorpius and Sagittarius, lanes which widen into the Great Rift northward toward Cygnus. To photograph this galactic structure, you will get amazing results with the new high speed films. Use your wide angle lens, set at infinity and wide open (my 28mm at f/2.8 does quite well) and either Fuji 1600 or Konica 3200 and time exposures of about a minute to avoid trailing. Take a mosaic from southern to northern horizons, and if you stay up all night, try to extend your mapping into the fall and winter portions rising later in the night. When you get over just how well your photos turned out, tape them together with some overlap to show your class their own map of the galaxy. Blow up the best negatives for an even more imposing bulletin board or permanent display.

Pay particular attention to the colors. Everywhere along the plane of the galaxy you will find red H II regions; we have already mentioned M-8, the bright Lagoon Nebula of Sagittarius, and M-42 of Orion. Both stand out like sore thumbs, but there are plenty of other H II regions which show up much better on your photos than visually--have your students identify them in your star atlas. Note also the stellar populations; the photos of Sagittarius will show a yellowish stellar background as you look downtown; this is population II, the older stars now dominated by aging evolved giants, now bloated and cooling in the final stages of their evolution.

But farther north, the galaxy as a whole takes on a bluer appearance, with the disk dominated by young population I stars still being born in the H II regions. This is especially true in the directions of Orion (looking behind us where the solar system has just passed) and Cygnus (ahead of us as we revolve around the center of the galaxy).

If you go south, by all means take along your camera, cable release, fast film, and tripod. The brightest portions of the Milky Way lie in the southern sky. Photograph the galaxy near the Southern Cross of spectacular emission and dark nebulas, and very colorful clusters. The eta Carinae Nebula, just west of Crux, is easily seen with the naked eye, and is really brighter and more complex than even the famed M-42 in Orion. You can get a good bit of structure and dust lanes in it with even a 135mm telephoto. It houses eta Carinae, probably the most luminous star in the Galaxy (estimated at over a million suns!) and a good supernova candidate. About 150 years ago, this variable star was even brighter than Canopus, although it has now faded to sixth magnitude. And of course if you are far enough south, observe and photograph those companions of the Milky Way, the Magellanic Clouds.

188

Dark Constellations of the Inca

To us and most other cultures, constellations are bright patterns of stars. But the Milky Way is much brighter south of the equator; a student of mine who grew up in the Outback of Australia tells me of seeing it cast his shadow when Sagittarius was at the zenith on a dark clear desert night. For the Incas, the dark dust clouds silhouetted in front of most of the galaxy were the basis of about twenty dark constellations. In their imaginations, they picked out a dark outline of a llama, a frog, a beaver, and several other creatures to add to the celestial zoo. Look at your long exposure photos of the Milky Way and invent you own dark constellations in the shapes of the dark nebula, such as the Coalsack just to the southeast of Crux, or the Great Rift running through Cygnus and Aquila. Of course, our popular culture has already done this on a limited scale in Orion, with the famed Horsehead Nebula. For more dark nebulas, read "Bardnard's Magnificent Milky Way," in Astronomy for June 1996, page 32.

Stellar Associations

In most cases, the constellations are random groupings of unrelated stars at a variety of distances and luminosities. But in the case of four of the most prominent constellations, this is not entirely true. These are stars arrayed too loosely to be called clusters, but still of about the same age, moving together with common parallel motions in space; they are the stars formed along the leading edge of the spiral arms.

In the Big Dipper, the middle five stars are moving together through space; this Ursa major moving group may also include Sirius, on the other side of the sky. If so, then our solar system lies within this loose association at present, but is not a part of it; instead we are just "moving through the neighborhood".

The bright blue stars of the two brightest constellations, Orion and Scorpius, are also members of associations of type O and B stars. It is notable that the brightest two red supergiants, Betelgeuse and Antares, happen to lie in these constellations; in both cases, they are not part of the associations, and actually lie in front of the much younger groups of even more luminous blue stars.

A good bit more distant, the Perseus Association actually lies in the arm of the galaxy beyond our own, but the bright blue stars of Perseus are some of the most luminous in the galaxy. Imagine how much brighter the Perseus Association would appear if it were closer. Many of these O and B stars have absolute magnitudes of -7 to -9, meaning that if they were about 30 light years away (the distance to Vega, for instance), they would be easily visible in broad daylight; as they are going to evolve rapidly into supernova candidates, perhaps its just as well they are not so close...

How is an association different from an open cluster, such as the Pleiades? Associations are much larger, spanning hundreds of light years in some cases, while a typical open cluster is only about ten light years across. Since the associations are highlighted by very luminous and thus short-lived OB stars, the associations disappear quickly, in just a few million years in most cases. Their brightest members die as supernovae, and the less massive stars interact with other stars and clusters in the thickly populated disk to blend into the background; our stellar "formation" has lost its leaders, and the privates all go AWOL. This is happening to the Ursa Major Moving Group, which could not be identified as an association unless we were actually imbedded inside it at present.

189

ANSWERS TO STUDY EXERCISES

1. The absorption and scattering of light by interstellar dust limits the range of optical investigations within the galactic disk. The most distant objects that can be studied within the place of the Milky Way by an optical telescope are only 3,000 light years away. Radio and infrared telescopes can probe far more deeply, as these waves are not blocked by the dust grains.

2. Population I Cepheids are good spiral arm tracers because they are very luminous and so can be seen over large distances. Population I stars are found in spiral arms. Also, the period luminosity relation for Cepheids makes it easy to find their distances.

3. The 21-cm radio observations only give the line of sight velocities of hydrogen gas clouds. Knowing the velocity of the sun around the galaxy enables astronomers to calculate the orbital velocities of the gas cloud, if they can translate a radial into an orbital velocity. The rotation curve allows them to do this.

4. Spiral arms cannot be material arms because they must be recently formed; the arms contain many bright blue supergiants (O and B associations) of very short life spans. Also, if the arm were a material object, it would have become twisted around the core of the Galaxy after many rotations, losing its form and identity over time.

5. Spiral arms contain material typical of starbirth, such as H II regions, Population I stars, OB associations, H I regions, and molecular clouds.

6. The density-wave model accounts for the persistence of the spiral arms as well as structural features such as the abundance of young stars and dust clouds. It also explains the grand layout of two or four major spiral arms. However, it is inadequate in that it does not account for the origin of the waves, nor does it provide a mechanism to prevent the waves from dissipating.

7. Using the sun's orbital period of about 250 million years, and its distance of about 30,000 l.y. from the nucleus, we get about 100 billion solar masses lying between us and the center. But if we use the rotation curve on objects beyond the sun's orbit toward the edge of the Galaxy, we do not find the expected decline in orbital velocities, as Kepler's Laws predicted. Instead, the rotation curve flattens out, indicating the majority of the mass must lie still farther out.

8. Globular clusters orbit in the spherical halo around the Galaxy's core. They contain metal deficient Population II stars in varying abundances. This implies the stars in the globulars were born when the metal abundances of the interstellar medium was low, early enough that few supernovas had occurred and few red giants had developed stellar winds. Thus they formed early in the history of the Galaxy, before the disk of the Galaxy was established by its rotation.

190

9. Some radio energy is produced by the synchrotron process, so the nucleus must contain high-speed electrons (and a means to accelerate them, as well as a supply of electrons) spiraling along the magnetic field lines.

10. The rotation curve of the very innermost parts of the Galaxy, within a few light years of the core, allow us to calculate the mass as several million solar masses. Since this region is so small, the most reasonable form for this mass is a collapsed black hole.

11. The galactic center is about 28,000 light years distant, according to the distances to the globular clusters and their distribution. This is uncertain, with estimates ranging from 24,000 to 33,000 light years.

12. The older the Galaxy, the more supernova explosion will shower the disk with heavier elements. As the Galaxy ages, there is less hydrogen and more heavy elements in the ISM.

13. The orbits of the globular clusters are all believed to have the Galactic nucleus as one focus of their elliptical orbits.

14. These molecules are made of heavy elements, so the greater the proportion of these heavy elements in the future, the more dominant these dark molecular clouds will be in the ISM.

ANSWERS TO PROBLEMS & ACTIVITIES

1. We adapt Kepler's Third Law to use solar masses, AU, and years, thus $P^2 = R^3 / 10^{12} = (150,000 \times 6.3 \times 10^4)^3 / 10^{12} = 8.4 \times 10^{17}$, so the period is 9.2×10^8 years.

2. Assuming our orbit is circular, and that we are 28,000 light years from the nucleus, and that we are travelling at 220 km/sec, then the period $= 2 \times 3.14 \times 2.66 \times 10^{17}$ ly / 220 km/sec $= 7.6 \times 10^{15}$ sec, or 240 million years.

3. The orbital velocity $= 2 \times 3.14 \times R / P$, so it takes the material $6.28 \times 3 \times 9.5 \times 10^{12} / 100 = 1.8 \times 10^{12}$ sec, or 56,700 years to orbit the nucleus. The mass inside this radius must therefore be $M = 39.5 \times (2.85 \times 10^{16}$ m$)^3 / 6.67 \times 10^{-11} \times (7.7 \times 10^{11})^2 = 4.23 \times 10^{36}$ kg. or 2,2 million solar masses, all within 3 light years of the nucleus.

4. Sin x = 1/28,000, so from 28,000 light years, a jet a light year long would be 7.4 arc seconds.

5. If the globular's maximum distance is 350,000 ly, then its average distance must be less than 175,000 light years, or 1.66×10^{21} km. Plugging in a total mass of 7.2×10^{11} solar masses, or 1.37×10^{42} kg., we get $P^2 = 39.5 \times (1.66 \times 10^{21})^3 / 6.67 \times 10^{-11} \times 1.37 \times 10^{42} = 1.97 \times 10^{33}$, so the period is 4.44×10^{16} seconds, or about 1.4 billion years.

CHAPTER 19

The Universe of Galaxies

CHAPTER OUTLINE

Central Question: What is the structure and content of galaxies, and how are they distributed throughout the universe?

19.7   Superclusters and Voids
        A. The cosmic tapestry

19.8   Intergalactic Medium and Dark Matter

Enrichment Focus 19.1: Making Your Own Hubble Plot

CHAPTER OVERVIEW
Teacher's Notes

The material in this chapter is descriptive and fairly straightforward, and also tends to be mind boggling. Emphasize distance-measuring techniques; note that the methods outlined there utilize bootstrap methods similar to those for measuring stellar distances. You might want to have the class lay out a scale of distances to the Virgo cluster.

Slides will help illustrate types of galaxies. Hale Observatories has many slides of galaxies and clusters of galaxies in their catalog; the galaxies are identified by type. A laboratory exercise might involve classifying and counting galaxies in photographs of clusters.

Be aware that the large-scale structure of the universe seen to date is based on very limited samples of the sky. These may not be typical of the grand scheme, but they do give us an inkling. Be sure to emphasize that superclusters are fossils of the early universe, and so provide us with information about that time if we have physical models to explain the structures.

19.1 The Shapley-Curtis Debate

The visible universe, for as far as our scopes can reach, is dominated by galaxies. They come in a host of sizes, shapes, masses, and luminosities. With color photography, they reveal themselves as among the most beautiful objects in the cosmos. While the planets and stars stayed well out of each other's way, galaxies mix it up quite a bit, with evolutionary implications just now being unraveled. They, and questions about their evolution, are what spur us to the construction of the Hubble Space Telescope, the Keck telescopes, and the other new technologies which allow us to view farther out into space, and back into time.

To understand why Shapley defended the incorrect placement of spiral nebulas inside our own galaxy, you must realize that Shapley's new dimensions suddenly made the Milky Way 10x larger than previously expected. The human brain can expand its horizons only so rapidly; the new Milky Way seemed so huge as to encompass the entire cosmos. Also the astrometric astronomer Van Maanen believed he found evidence of rapid rotation in the spiral arms of the nebulas; if so, Einstein's limit on the speed of light implied these rapidly spinning arms could not be millions of light years distant, as Curtis proposed. Later van Maanen's data was proven incorrect, and Shapley dropped this objection and accepted that the Milky Way was but one of millions of "island universes," as Emmanuel Kant had suggested two centuries before.

Hubble's discovery of cepheids in the Andromeda Galaxy in 1925 first established the distance scale to these spiral nebulas was truly millions of light years, well beyond even Shapley's larger dimensions for the Milky Way itself. Shapley would agree to the change within a decade. Even great scientists make mistakes, and greater ones accept them and move on.

19.2 Normal Galaxies: A Galactic Zoo

The number that follows the letter E is giving us the eccentricity of the oval; remember an eccentricity of 0 is a circle, so E0 galaxies appear circular. E7 is the largest number because as rotation increases and the disk flattens, we find spiral arms growing if the galaxy is flatter than this. The S0 galaxies are believed to be the transitions between spirals and ellipticals.

Exactly why about a third of the spirals exhibit the bars is unknown. Our own Milky Way is now classified by some as SBb. Color photos reveal the bars to be made of yellowish population II stars; how they make such a sudden transition to the blue spiral arms which seem to abruptly break off the bars is a good mystery, one of many we find in examining galaxies.

Chemically the irregulars are more like the spiral arms but without their structure. The blue stars and H II regions make many think the smaller irregulars condensed later than the larger galaxies; perhaps as with clusters of stars, the most massive ones collapse first, and their formation governs in part the formation of the smaller members who come later. See the article on M-33 in the September 1996 issue of Astronomy on page 28.

Zeilik mentions the luminosity classes for galaxies, but I fail to find where he would place our home Milky Way. My personal guess is it is an average large spiral, probably in class III. The "Ghost Galaxies" article in June 1996 Astronomy on page 40 suggests more classes are needed.

It is notable just how confusing the galactic census can be; sort of reminds you of our problems with surveying the naked eye vs. telescopic stars, doesn't it? As spirals contain bright population I stars in their arms, they are quite luminous and visible over large distances. But the richest clusters (such as the Virgo Supercluster) seem dominated by ellipticals (more on that later); those little irregulars may be, despite their faintness, the most abundant type of galaxy after all, just as there are a lot more red dwarfs than even stars as bright as the sun.

19.3 Surveying the Universe of Galaxies

Our problems with determining the distances to galaxies, and from that the age of the universe, all stem from unreliable distance indicators. To be seen at millions of light years, an object must be very luminous; as our study of stars showed earlier, the more massive and luminous the star, the faster it evolves and more variable its energy output. While I think we can trust cepheid variables somewhat, the big red or blue supergiants are not nearly so dependable.

Some astronomers like H II regions, but I see again too much variation from galaxy to galaxy to depend on them. For instance, the Tarantula Nebula of the Large Magellanic Cloud is much larger than any H II region known in the much larger Milky Way galaxy. Zeilik does not include either planetary nebula or globular clusters in this listing on page 440, but both are thought to be more promising for distance indicators in the future. Likewise both supernovae 1987A and 1993J showed us that supernovae are not as predictable in their greatest luminosity as we had previously hoped. As Zeilik just noted, even the spiral galaxies differ in their luminosities; obviously mistaking a luminosity class I spiral for a II could throw the distance to the whole cluster of galaxies off by 20-30% So when one astronomer tells us M-51 is 20 million light years distant, and another says its only 15 million light years away, that's just the way it goes. It all depends on which objects you trust as your measuring rods. For more on interpreting the newest Hubble results, see "A Galaxy of News" in the June 1995 issue of Astronomy on page 40.

19.4   Hubble's Law and Distances

The first hint of the Big Bang came in 1905, when V. M. Slipher of Lowell Observatory obtained the doppler shifts of about 20 spiral nebulas. This is two decades before the Shapley Curtis debate, and most astronomers considered them to be condensing solar systems. As such nearby objects, we would have expected them to show random shifts, some approaching and blue shifted, others receding and red-shifted. But except for M-31 and M-33, Slipher found all the others receding from us, as if the Milky Way were repelling them. Some of the recession velocities were also larger than any observed for stars within our home galaxy, suggesting these objects were even farther away. But the next year, Percival Lowell died, and Slipher's career emphasis became serving as the great expert on the planet Mars, thus missing out on his chance to become the father of the Big Bang.

Hubble later continued Slipher's spectroscopic observations of galaxies, and found that they were not only red shifted, but that there appeared to be a correlation between how fast the galaxy was receding from us and how distant it was. This is, of course, called Hubble's Law.

Note as mentioned above that if you do not know the exact distances to galaxies, you can't calibrate Hubble's constant well, hence the factor of two in the accepted range of H, between 15 and 30 kilometers/sec/Million light years.

For another sample problem, consider a galaxy with a spectral line observed at 6,060 A and the rest wavelength of the line is 6,000 A; then we get a red shift of 60/6,000, or 1% of the speed of light. If we use $H = 15$ km/sec/Mly, then we get $D = v/H = 3,000$ km/sec / 15 km/sec/Mly, or a distance to this galaxy of 200 million light years; had we used $H = 30$ instead, the distance would have been halved, to only 100 million light years.

19.5   General Characteristics of Galaxies

We find almost as wide a range in the luminosities of galaxies as we did with stars, and an even wider range in galactic masses than with stars. It is worth remembering the differences in gas content of the galaxies influence greatly the M/L ratio. A spiral with a lot of gas and dust will still be producing hot O-B associations and appear much brighter than an elliptical with the same mass of older population II stars. Irregulars with even more star birth occurring will be even brighter per unit mass than spirals like Andromeda, where starbirth proceeds more slowly.

19.6   Clusters of Galaxies

Think of the Local Group as a small lagoon housing two large sharks, the Milky Way and the Andromeda Galaxy. Most of the other fish in the lagoon are remora which closely accompany the two big fellows; only a few fish are found out on their own. Let me illustrate.

Two of the larger fish in the Local Group may already have jumped out of our pond. The article, "Intruder Galaxies", in November 1993 Astronomy on page 28 suggests this.

Another point worth noting is that rich clusters of galaxies lack spirals, particularly in their centers. The closer the galaxies are, the more likely that interactions and collisions will strip the gas and dust from the spiral arms. When two spirals collide, the result may be two ellipticals, stripped of gas and dust, emerging from the collision. The gas and dust either quickly formed new stars during the collision, or if hot enough, was left behind as an X-ray emitting cloud spreading throughout the whole cluster, as described in the next section. Perhaps collision with the Sagittarius companion galaxy will create disruptions violent enough to glow much gas into the Local Group.

The small ellipticals in larger clusters may be future victims of galactic cannibalism. Located at the heart of the Virgo Supercluster, the monster galaxy M-87 appears to be gravitationally pulling other galaxies toward the black hole in its center. In the next chapter, we will see that such a huge appetite creates indigestion, for there are also energetic jets of matter being ejected from the nucleus of M-87. So in studying interactions, some get eaten, while activity may in turn lead to the creation of others; the Magellanic Clouds for instance, contain blue globular clusters so they must be only half the age of the Milky Way itself.

19.7 Superclusters and Voids

For a good example of voids, consider galaxy clusters like soap bubbles. The galaxies lie along the soap film, and congregate in superclusters where bubble surfaces meet; but most of the space is empty, like the voids Zeilik describes on page 454. Or is it? The "Ghost Galaxies" article suggests there is much matter there, but forming galaxies at a far slower pace, hence still dark.

19.8 Intergalactic Medium and Dark Matter

In considering the dark matter problem, another bit of evidence lies in the Viral Theorem. If we consider the mass necessary to hold a cluster of objects together gravitationally, it must be greater than half the average kinetic energy of the cluster members, or the cluster will fly apart, much as our own solar system escaped from its parent cluster billions of years ago. Typically open clusters do not contain enough mass to achieve such stability, while the far older globular clusters are much richer in stars and will stay gravitationally bound for all time.

But what about clusters of galaxies? If we use their doppler shifts to find their motions within the cluster, the random motions appear to be too great, and the clusters must fall apart. This is just based on the mass seen as visible stars within the galaxies, however. For as Zeilik notes, clusters are the rule--bachelor galaxies are far more rare than single stars. If anything, the great clusters of galaxies are still condensing. Again, there must be much material in the form of dark matter to account for this gravitational stability of the clusters of galaxies.

As already mentioned, the collisions and interactions of galaxies in rich clusters can heat up the intergalactic medium, creating the X-ray emitting regions. But there may be a lot of cold hydrogen out there, too. Some believe the voids are not truly empty but that the conditions for galaxy creation just take longer there. There are controversial radio observations of what may be "dark galaxies", great clouds of hydrogen and helium already forming swirling disks in space, yet not yet condensing into actual luminous stars. Meanwhile the supercolliders at CERN in Switzerland or in Batavia, Illinois may turn up tiny, supermassive subatomic particles which may also play a major role in the problem of dark matter. As noted in the November 1993 Sky & Telescope on page 9, the MACHOs (brown dwarfs) may have just been found via gravitational lensing. Keep an eye on the journals for new developments in both galactic astronomy and particle physics to solve this mystery.

ADDITIONAL RESOURCES

Slide 19.26 is a new HST image of the core of M-32. Transparencies 19.2, 19.3, 19.19ab, 19.23, 19.34, and 19.35 also update your class presentations.

The first episode of the Astronomers series, "Where is the Rest of the Universe", fits well here or in chapter 18.

SUPPLEMENTARY ARTICLES

1. "Discovery of the Expanding Universe," Astronomy, February 1985, page 18.
2. "Forces that Shape Spiral Galaxies," Astronomy, December 1987, page 6.
3. "Search for the Most Distant," Astronomy, June 1988, page 20.
4. "A New Yardstick for the Universe," Astronomy, November 1988, page 60.
5. "The Farthest Galaxies," Astronomy, December 1988, page 56.
6. "Collisions between Galaxies," Astronomy, May 1989, page 44.
7. "How Far to the Galaxies?," Astronomy, June 1989, page 48.
8. "The Legacy of Edwin Hubble," Astronomy, December 1989, page 38.
9. "A Grand Gathering of Galaxies," Astronomy, March 1991, page 39.
10. "When Galaxies Go Wrong," Astronomy, October 1991, page 74.
11. "Going Deep for Galaxies," Astronomy, March 1992, page 68.
12. "Galactic Archeology," Astronomy, July 1992, page 28.
13. "Odd Couples--Galaxies in Collision," Astronomy, November 1992, page 36.
14. "Counting to the Edge of the Universe," Astronomy, April 1993, page 38.
15. "Exploring the Virgo Cluster of Galaxies," Astronomy, April 1993, page 56.
16. "The Great Attractor," Astronomy, July 1993, page 40.
17. "What Puts the Spiral in Spiral Galaxies," Astronomy, September 1993, page 34.
18. "Intruder Galaxies--Violence in the Neighborhood," Astronomy, November 1993, page 28.
19. "Eyeing the Local Group," Astronomy, November 1993, page 94.
20. "Our New! Improved Cluster of Galaxies," Astronomy, February 1994, page 26.
21. "The Seven Mysteries of Galaxies," Astronomy, March 1994, page 38.
22. "The Debut of Galaxies," Astronomy, October 1994, page 44.
23. "The Galaxy Time Machine," Astronomy, March 1995, page 44.
24. "A Galaxy of News," Astronomy, June 1995, page 40.
25. "Our Strange, Scrappy Ancestors," Astronomy, December 1995, page 52.
26. "A Window to the Deep," Astronomy, April 1996, page 82.
27. "Ghost Galaxies of the Cosmos," Astronomy, June 1996, page 40.
28. "Galaxies of Autumn," Astronomy, June 1996, page 70.
29. "Is M-33 Younger than the Milky Way?," Astronomy, September 1996, page 28.

DEMONSTRATIONS AND OBSERVATIONS

Observing Other Galaxies

Due to the tremendous distances involved, most other galaxies are far too faint to be easily observed with small telescopes. But if you can get a fairly dark site, there are a few bright enough to make the effort with a telescope 6" or larger.

Under clear, dark autumn skies, the Andromeda galaxy, M-31, is faintly visible as an oval blur just north of beta (the middle star in the body of) Andromeda, about a third of the way to her mother Cassiopeia. With binoculars it stands out well, as does the fainter face-on spiral M-33 a few degrees south of it in Triangulum. With a 4" scope at 20x, note the two fainter elliptical companions of the Andromeda spiral, one compact and almost in front of the bigger galaxy, the other more diffuse and just above the spiral; they are M-32 and M-110, respectively.

A 10" or larger scope under really dark skies will show you some dust lines in M-31 and even some of its brightest globular and open cluster. Even more detail is visible in big scopes looking at M-33; this face on spiral reveals its arms well at 70x, and in them you can spot some the brightest H II regions, forming new stars.

In both cases, note the nucleus of each large spiral is a tiny, bright, slightly fuzzy object. Recently it was discovered the Andromeda Galaxy has a double nucleus, as discussed in the November, 1993 article in <u>Astronomy</u>. Apparently M-31 collided with two other galaxies and expelled them from the Local Group, and perhaps devoured one of their smaller companions in the process. You may be looking at the hot accretion disk around the black hole that holds each galaxy together. The Milky Way may gain a secondary nucleus as it consumes the Sagittarius companion.

Farther south, we look beyond our own Local Group to the nearby Sculptor Cluster of Galaxies. The best is NGC 253 which is easy in binoculars. Look for a nice globular, NGC 288, in the same binocular field, just south of beti Ceti. It shows a lot of dust lanes in 10".

As we look in the southern fall sky, we look away from the obscuring gas and dust of our home galaxy, and downward toward the south galactic pole. A star atlas will plot hundreds of fainter galaxies, within range of larger amateur telescopes.

Both winter and summer skies are dominated by our own Milky Way. Thus while there must be galaxies out there, many are hidden in the "Zone of Avoidance" along the plane of the Milky Way. The next good season for galaxy searching is Spring, when the northern galactic pole in Virgo rides high overhead.

Between the head of Ursa major and Polaris are a fine pair of galaxies, M-81 and M-82, both visible in binoculars. M-81 is a normal big spiral, comparable to the Milky Way; it was the site of a supernova visible in 4" telescopes in the spring of 1993. M-82 is an active, starburst galaxy like those featured in chapter 20. They look very nice together in the same 30x field with a 6" scope. Another nice pair are M-65 and M-66 in Leo, visible in scopes 4" and larger. As you can see, galaxies, even more than stars, seem to always come in clusters. This in turn means that they often collide, and the finest visual example lies just south of the end star in the Big Dipper's handle; M-51, the Whirlpool, is the product of the small elliptical to the bottom sideswiping the big spiral over the last several million years. This tidal interaction makes the spiral arms even more prominent that usual, and I have noted spiral structure in this show object in scopes as small as 4" under dark sky conditions. You can probably see the larger galaxy even in a 30mm finder scope or binoculars.

Farther south, in the arms of Virgo, lies the rich Virgo Supercluster. Here a 10" scope reveals galaxies all over the place, with as many as seven in the same 50x field of view in some places. The heart of this cluster is occupied by the giant elliptical, M-84 (figure 16.3), faintly visible in binoculars even though it is about 50 million light years distant. Visually more interesting is M-104, the Sombrero galaxy. The dust lane in this almost edge on spiral stands out well at 50x in the six inch or larger scope.

Much farther south in the autumn skies are the Magellanic Clouds, best seen from below the equator. They appear to the naked eye like small patches of the Milky way which broke off (and that may in fact be their origin). Binoculars reveal the largest stellar nursery yet found, the Tarantula Nebula, on the edge of the Large Magellanic Cloud; larger telescopes focused on it reveal a wealth of details in this much larger and brighter but 100x more distant version of the Orion Nebula. It likely is the mother of Supernova 1987A, which lay just a few degrees away.

Modeling the Local Group and Beyond

If you have one of those 15" Milky Way plastic models, use it as the basis for a class demonstration of the scale of intergalactic distances. If you have two, even better--let one of them stand for the Andromeda galaxy, only slightly larger than the Milky Way. If not, make a 16" wide cardboard spiral galaxy to represent M-31, and place it about 30 Milky Way diameters distant, or about 30 feet away. Add a 1x3" styrofoam oval on the far edge of the Milky Way to represent the Sagittarius companion found in 1995. Use two irregular clumps of old styrofoam, about 3" and 2" across, to represent the Magellanic Clouds. Hang them about 20 and 25 inches below your Milky Way model itself. M-33 in Triangulum is a spiral cardboard disk about 8" across, and in the general direction of Andromeda, but about 35 feet distant from the Milky Way.

For a sense of scale, consider our small cluster of galaxies as a part of the much larger Virgo Supercluster. Our scale in the Local Group was that a million light years is about ten feet, so the Virgo Supercluster, whose center new Hubble data indicates is about 65 million light years distant, would lie about two football fields distant, and encompass an area the size of a large high school gym.

If the known universe is ten billion years old, as the latest Hubble data seems to indicate, then it would be scaled to cover an area of 20 miles, or about the size of a moderately large city like Pensacola. Even the larger versions with a smaller value of H would be no larger than the biggest cities.

ANSWERS TO STUDY EXERCISES

1. Elliptical galaxies appear redder than spirals. Ellipticals are made almost entirely of Population II stars, such as in the globular clusters. The gas and dust in the disk of spirals gives birth to bright blue stars in the arms of spirals; ellipticals lack that gas and replacement method.

2. While spirals do contain much gas and dust in their disks, smaller irregulars contain an even greater fraction of gas and dust throughout. The cores and halos of spirals are poor in gas and dust, but irregulars are rich in gas and dust throughout.

3. Irregulars are far smaller and less massive than spirals; apparently if a galaxy has a mass of several billion suns and much gas and dust, spiral forms are easily maintained.

4. We can resolve individual supergiant stars in nearby galaxies out to about fifteen million light years. Also the spectrum of most distant galaxies appears similar to the Milky Way's, suggest we are seeing the combined spectra of many billions of stars.

5. All distances will be doubled if H is halved, since $D = v/H$.

6. Cepheid luminosities increase as the period of pulsation lengthens. If we know what type of Cepheid we are observing and can watch it long enough to find its pulsation period, we then use the period-luminosity relationship to estimate the star's luminosity. By comparing this with the star's actual brightness, we can calculate just how distant the galaxy must be for its Cepheid to appear this faint, based on the observed flux and the inverse square law.

7. Intergalactic gas has not been detected by using methods that could find dust, neutral hydrogen, or even cool, ionized hydrogen.

8. Basically, a supercluster is a long string (filament) of galaxies. They are NOT spherical in shape, as was thought in the 1960s.

9. We see depth by measuring the red shifts of galaxies in the clusters. From their doppler shifts, we know how fast the cluster is receding, and IF we know H well, then we can estimate the distances to the superclusters.

10. In just our neighborhood the least massive irregulars are also the most common, just as the most common stars are the least massive. If we takes our region to be normal, this will probably be typical of the cosmos as a whole. Distant clusters often appear dominated by large, bright spirals, but at their distances, we could not even see the more numerous but far smaller irregulars.

11. The Virial Theorem lets us compare the random motions of member galaxies in a cluster with the total gravitational potential energy needed to hold the cluster together. If only the luminous mass of the cluster were considered, the kinetic energy of the galaxies would long ago have broken the clusters apart. Yet the clusters hang together--single galaxies are very hard to find. We must conclude that much dark matter lies in the halos to hold the clusters together.

12. The flat rotation curve in the outer regions of the spirals again implies much dark matter must be present farther out, to keep the speeds high; if the distribution of mass were similar to the solar system, with most mass toward the center, the outer regions would slow down considerably, in accord with Kepler's laws. It is estimated our Milky Way has ten times more matter out in the halo than in the visible disk.

13. There is no detectable reddening in the light of distant galaxies due to the dust between them and us, unlike the considerable reddening of distant stars in our own Galaxy due to dust concentrated in our galactic plane.

14. Based on sizes, the smaller galaxy is four times as distant. If they are equal in luminosity, the inverse square law tells us the smaller galaxy is sixteen times fainter.

ANSWERS TO PROBLEMS & ACTIVITIES

1. If the redshift is .10, then it is receding at .10c, or 30,000 km/sec. Using $H = 20$ km/sec/Mly, we get $D = v/H = 30,000/20 = 1.5$ billion light years.

2. The Local Group has a radius of about 1.5 million light years. As the Andromeda Galaxy and our Milky Way are the vast majority of the mass of the Local Group, we estimate their combined masses is 2×10^{12} solar masses. Using $P^2 = 39.5 \, (1.425 \times 10^{22} \text{ m})^3 / 6.67 \times 10^{11} \times 3.8 \times 10^{42} = 4.5 \times 10^{35}$ sec.^{2}, so the period of revolution is 6.7×10^{17} seconds or 21.25 billion years; so this small galaxy has not yet completed a single revolution since the Big Bang.

3. Wien's Law gives peak wavelength $= 2.9 \times 10\text{-}3 / 10^6$ K $= 2.9 \times 10^{-9}$ m or 29 Angstroms, in the X-ray portion of the spectrum.

4. Again using the method in # 2, we assume the two similar galaxies are comparable in mass, so the barycenter of the pair lies 1.1 million light years from each, and the combined masses again are 3.8×10^{42} kilograms, we get $P^2 = 39.5 \times (1.045 \times 10^{22})^3 / 6.67 \times 10_{-11} \times 3.8 \times 10^{42} = 1.78 \times 10^{35}$ sec.^{2}, so the pair's period of revolution is 4.2×10^{17} sec. or 13.4 billion years.

5. If the diameter of a supercluster is about 250 million light years, and a member galaxy is moving at 300 km/second, then the galaxy moves 9.5×10^9 km/year, or about .001 ly per year. This means the galaxy would take 250 million ly/ .001 ly/year = 250 billion years to move all the way across its supercluster.

6. If we use $H = 20$ km/sec/Mly, we get $D = 6650/20 = 332.5$ million light years.

7. Sin 4.5 deg = .0785, so M-31 is 12.75 times more distant than its diameter. If it is estimated to be 2.2 million ly distant, this means it is 173,000 light years across, a little larger than the Milky Way.

8. Using the Milky Way's diameter of 100,000 ly as average for spirals, if this new galaxy is 10' across, sin 10' = .00291, so it is 344 times more distant, or about 34 million ly distant.

201

CHAPTER 20

Cosmic Violence

CHAPTER OUTLINE

Central Question: What observational evidence do we have for violent activity in objects beyond the Milky Way Galaxy?

CHAPTER OVERVIEW

Teacher's Notes

Quasars, like pulsars and black holes, have captured the imagination of the public and are certainly one of the popular mysteries in the universe. Point out that while we don't have a clear idea of the nature of quasars, astronomers are attempting to reconcile models, based on physical principles, with observations. This approach is the same process that was emphasized in the first seven chapters. Our understanding of quasars at this point is not very different from our understanding of the solar system in the fifteenth and sixteenth centuries. It may help to define some terms more completely than is done in the text. The terms thermal and nonthermal radiation have definitions in the body of the text. Point out that synchrotron radiation is only one form of nonthermal radiation--but most common astronomically.

The idea that the period of variations in the light output of quasars gives an indication of the size of the region that is varying is a new one, and an important one to many areas of physics and astronomy--determining sizes of variable stars or disturbances on the sun or in the earth's atmosphere. A demonstration of this principle can be done by an acoustical analogy, using two speakers emitting the same signal, but with one of them having a delay line between it and the amplifier or sound source. The delay may also be inserted on an audio tape by separately recording music on two stereo tracks of the tape, one track out of synchronization with the other by various times, ranging from a fraction of a second to a few seconds. If done carefully, you could show that for a given time lag between tracks (analogous to a given size emitting region), music with a slower tempo should sound clearer or better than faster music.

A strong argument that favors the association of quasars with galaxies is that some quasars are seen with galaxies nearby, more than would be expected from chance alignments in space. This argument would be stronger if a physical connection between quasars and galaxies, such as bridges of gas and dust, were confirmed. If quasars were shot from active galaxies, we should expect some of them would be shot in our direction, and they would then blow blue-shifted spectral lines. However, no blue-shifted quasars have been found. And no definite evidence has been found to a galaxy nearby by a bridge of matter.

20.1   Violent Activity in Galaxies

While most galaxies near use appear fairly placid, at greater distances we are looking back farther into time, to the stages when galaxies were young and more active and energetic. While these active galaxies and quasars take on a host of appearances to us, it may be that we are just viewing different orientations of the same basic process in their core.

The energy of the active galaxies comes from their nuclei, hence the AGN designation. As with the Crab Nebula, much of the energy is synchrotron radiation, polarized and associated with very strong magnetic fields. As with pulsars, the energy output comes from a very small region and is in many cases quite variable over short time periods.

An accretion disk around a black hole is suggested by the presence of strong bipolar jets in many cases. Such jets are found on a far smaller scale around YSOs in Chapter 15 and around SS-443 and other collapsed stellar remnants in Chapter 17; see the Hubble supplement in the text for very striking images of bipolar jets in a range of sizes and energies.

The jetlike structures in Figure 20.1 indicate that the Milky Way has had a violent past, with a more active nucleus than now. The new discovery that the Andromeda Galaxy has a binary nuclei indicates that it too was once more active than at present; the Milky Way seems to be forming stars faster than its larger and more sedate sister galaxy. If starlight is the only source of visible energy in the galaxy's spectrum, we should see an absorption spectrum, with the lines fuzzy from the motions of the member stars. If we note as well bright emission lines, these hint a strong synchrotron source is ionizing hot gases near the nucleus and adding these bright lines to the spectra of AGNs.

20.2   Radio Galaxies

Bipolar jets help distinguish the compact and extended radio galaxies. Compact radio galaxies have the source coinciding with the optical nucleus. The extended radio galaxies are where the jets have spread the expelled gas over a large region; in most cases this is observed as dual lobes of radio emission on opposite sides of the accretion disk and perpendicular to it. Some of these radio lobes extend over millions of light years and make these galaxies the largest single objects yet found.

We briefly discussed M-87 in the previous chapter. Do not expect to see any of this violence in smaller telescopes, however. A curiosity at M-87 and with several quasars and other AGNs is that we only observe the jet coming out in one direction, not bipolar as our model predicts. Another is that the amount of material pumped out in the seven knots observed in this jet varies in short time periods of just a few months; the active region around the nucleus can not, due to the limitations of the speed of light, be much larger in the inner Oort Cloud, just a few light months in diameter.

Since Cygnus A is (after our own Galaxy's nucleus) the strongest radio source in the whole sky, optical attention was focused upon it early on. The first photos seemed to reveal two galaxies in collision. Later images showed it was one giant elliptical, cut by a dark dust lane much like NGC 5128, much closer to home. Centaurus A is perhaps only about 10 million light years away and thus much easier to observe than still more energetic but much more distant Cygnus A. Note how thin the jet in Figure 20.6 is, compared to the huge extended radio lobes.

20.3   Seyfert Galaxies and BL Lacertae Objects

For Seyferts, first note the color of the nucleus; it is bright blue, compared to the yellow we expected for normal spiral nuclei. Next contrast the spectrum. In a normal galaxy, we would observe the combined light of their stars as fuzzy absorption lines; in a Seyfert type spiral, the bright nucleus give us instead emission lines, typical of a hot thin gas. The broadness of these lines indicates a lot of localized doppler shifts, from very high velocities in the gas cloud.

Again, the nuclei appear quite small, only a few light years across. The relationship with close binary galaxies suggests some sort of tidal interaction spurs the short-lived Seyfert phase, for only about 1-2% of all observed spirals are classified as Seyferts.

BL Lacertae objects do not show emission lines typical of other AGN and show the fastest variation in their energy output of any AGN, in some cases within ten minutes. They may be at the core of elliptical galaxies, based on the fuzziness around the visible point of light. There is also the possibility that these galaxies do not have enough gas near the active nucleus to become ionized and create the Seyfert's emission lines. As discussed in chapter 19, the gas and dust content of individual galaxies can vary greatly; perhaps going back to the very origins of the galaxies. See "Our Strange, Scrappy Ancestors," in Astronomy for December 1995, page 52.

20.4   Quasars: Unraveling the Mystery

Quasar stands for "quasi stellar radio source". When first cataloged at Cambridge, they appeared to be optical points of light, or stars. But most stars, like the sun, are rather weak radio sources; what made these stars different? When we turned our large scopes toward them, their spectra gave us two distinct surprises. First we found narrow, bright emission lines, not the normal broad absorption lines. Even more perplexing, these lines did not seem to correspond to any known elements. Were these strange objects made of a completely alien chemistry from that of the rest of the universe? For several years in the early 1960s, this idea was taken very seriously as many astronomers and major observatories made little progress.

As vividly shown in "Beyond the Milky Way", Marteen Schmidt of Cal Tech was staring at the line spectrum of 3C 273 in 1963 when he suddenly realized the spacing of the three brightest emission lines lay in the same pattern as the three fainter balmer lines of hydrogen. Of course, people had looked for hydrogen to be present before; because the spectral lines were shifted an astounding 16 % of the speed of light, no one had recognized them. Remember the brightest line of hydrogen, the alpha line, is normally in the red; in this quasar, that line was red shifted all the way into the infrared, and hence optically invisible. The lines we normally see at blue-green and two more in the violet ended up in the yellow, green, and blue-green instead.

This created an immediate controversy. If the red shift gave a recession velocity of 48,000 kilometers per second, then an H value of 15 yielded a distance to the quasar of over three billion light years distant. Yet 3C 273 is visible in backyard telescopes as a 12th magnitude star in Virgo; a normal galaxy at that distance would be about magnitude 17, far beyond the range of a 6" telescope. If you trusted its red shift and Hubble's Law, this object was hundreds of times more luminous than any normal galaxy. Yet the image showed the energy emitting region was far smaller than any galaxy, and the variability of some quasars put it at only a few light months across.

Since then, many more distant quasars have been discovered. For most we must apply the relativistic correction, due to Einstein's limits on the speed of light and the strange things that happen to distance and times as v approached c. Here is a real example from current studies.

We observe the lyman alpha line of hydrogen normally in the ultraviolet, at about 2,000 angstroms. But for a faint quasar, our new technology reveals this line red shift all the way across the visible spectrum into the infrared, at 10,000 A. The conventional doppler shift formula would then tell us this object was receding at 8000/2000, or four times the speed of light. But Einstein's answer is that both time and distances are distorted as we approach the speed of light. The relativistic correction formula uses the red shift factor, Z, to stand for the apparent recession factor (or 4 for our example quasar above), and gives us the real velocity as $v/c = (Z+1)^2 - 1$ divided by $(Z+1)^2 + 1$; thus for our quasar, $(4+1)^2 -1 /(4+1)^2 + 1 = (25-1)/(25+1) = 24/26$, or just 93.2% of the speed of light--no problem after all. Still, this is a recession velocity of 277,000 kilometers per second; if we use H = 15 again, we get an apparent distance of 18.5 billion light years to this object. Thus we are seeing an object whose light was created when the universe was only about 7 % as old as it is today.

There may not be any quasars in the whole universe right now, despite the fact that our radio and optical telescopes are finding hundreds of them. Consider that the closest known quasar is over 2 billion light years distant; it was passing though the quasar stage two billion years ago and probably has evolved into a sedate, middle-class spiral by now.

If there were any closer full-scale quasars, they would stand out like sore thumbs; one in the Virgo supercluster, 50 million light years away, for instance, would be visible as a third magnitude star, easily seen with the naked eye. This implies that the quasar stage was associated with the violent youth and childhood of galaxies (perhaps even an juvenile delinquent stage) that all the galaxies around us have matured past. When the universe was younger and denser, there was more matter to feed the black holes at the cores of condensing galaxies; such well fed nuclei are the source of AGNs, and quasars are simply the most luminous phase of this phenomena.

Over time, most extra material close to the core is either swept up into the black hole, or condenses into stars in stable orbits beyond the accretion disk. As the core cools, we see the galaxy around the AGN as a Seyfert type spiral, no longer overshadowed by the AGN.

20.5 *Double Quasars and Gravitational Lenses*

Our new technology has now revealed numerous examples of how the bright images of distant quasars have been gravitationally lensed by the gravities of closer galaxies. The actual appearance of the image depends upon the symmetry of the quasar and galaxy; a perfect alignment can create not just multiple images, like the Einstein Cross, but even an arc or ring of quasar light around the intermediate galaxy. See the Hubble supplement for more of these images.

20.6 *Troubles with Quasars*

Not everyone accepts the vast distances and hence high luminosities of the quasar's red shifts as being accurate. Halton Arp in particular has challenged this interpretation, and in its place proposed the Local Hypothesis (not to be confused with the Local Group of Galaxies in the last chapter). He agrees the red shifts may be indicating real speeds, but does not attribute these high speeds to great distances, as Hubble's Law would imply. The high velocities are instead the product of great energy expelling "balls of fire" from active nuclei of nearby galaxies. In support of this, there are several cases where quasars seem to be aligned with jets in known active galaxies. The most famous case, Markanian 205, has a quasar seemingly linked to the nucleus of a spiral galaxy by a luminous bridge in the form of a jet from the galaxy's nucleus directly to the quasar. If this is true, the galaxy and quasar lie together in space. Yet the quasar has a red shift ten times larger than the galaxy's; Hubble would have you thus believe the quasar must be ten times more distant! Perhaps the Hubble Space Telescope, with high resolution, will prove the jet is real, but not connected at all to the quasar, as most scientists now believe.

A problem with the Local Hypothesis is that statistically not all such expulsions from nearby galactic nuclei could be seen receding; random orientations would make half of them approach us and show blue shifts. Searches of the entire sky over the last two decades have yet to turn up any quasar with a red shift smaller than .1c, much less any blue shift at all.

If we then accept the vast distances as being correct, this implies thousands of times more energy than our whole galaxy now radiates can come from a region only light months across. No massive superstar could do this; at their best, fusion reactions can convert only .007 of the mass of the fuel into energy. The generic model proposes a supermassive black hole will do a far better job. The hot accretion disk has infalling matter accelerated close to the speed of light. Not all matter falls into the black hole itself; if the Kerr model of a spinning black hole is correct, such an engine might even transfer 90 % of the infalling mass back out as energy, 150X more energy than from fusion.

The generic model implies that the more massive a black hole is, and the better it is being fed, the more energetic it is. Also its appearance to us depends upon our viewing angle, much as with eclipsing binary stars. Active radio galaxies are perhaps viewed from the side, with the active nucleus hidden (as with NGC 5128 below), while brighter BL Lacertae objects and quasars are seen face on, with the brilliant nucleus visible. Support for this model comes from observations of the very high speed of stars close to the nucleus of M-87, implying a very strong gravitational field, from a black hole of about five billions solar masses, is governing their motion. More recent results indicate similar, but less massive black holes exists in the Whirlpool Galaxy, M-51, and other galaxies. There may be a pair inside M-31 in Andromeda. Remember, these black holes may be gravitational glue necessary for any large galaxy to form.

Another way of attacking the Local Hypothesis is to find examples of quasars in clusters of galaxies. In that case, the quasar will be hundred of times brighter than the normal galaxies in the same cluster; but as all lie at a common distance, the quasar and surrounding cluster galaxies will all show similar red shifts. Numerous examples of this happening are now known, such as 3C 206 mentioned by Zeilik. One cluster recently studied even had two quasars in it.

Recently a type I supernova was observed on the edge of a quasar's parent galaxy. The brightness of the supernova was in good accord with the estimated distance to this quasar of about three billion light years.

The hyperactivity of quasars may tie in with galactic collisions and interactions. If two galaxies are colliding almost head-on, then sizable amounts of interstellar gas will be introduced to the central, normally gas poor nuclear regions. Again, a well-fed black hole is a tremendous energy engine; only feeding it one solar mass a year could power the AGN phenomena.

ADDITIONAL RESOURCES

The cover of Chapter 20 and 20.29 are both slides; also use transparencies 20.3, 20.12a through f, 20.13, 20.16, 20.21, 20.25, 20.31, and 20.32.

The second episode of The Astronomers series, "Searching for Black Holes" deals with the energy sources of quasars and active galaxies.

SUPPLEMENTARY ARTICLES
1. "M-87, the Monster at the Heart of the Virgo Cluster," Astronomy, May 1987, page 6.
2. "Chasing the Monster's Tail -- Cosmic Jets," Astronomy, August 1990, page 28.
3. "A Journey to the Center of the Galaxy," Astronomy, July 1991, page 74.
4. "A Quasar Lights Up the Universe," Astronomy, September 1991, page 42.
5. "When Galaxies Go Wrong," Astronomy, October 1991, page 74.
6. "The Emerging Picture of Quasars," Astronomy, December 1991, page 34.
7. "Odd Couples -- Galaxies in Collision," Astronomy, November 1992, page 36.
8. "Have Astronomers Solved the Quasar Enigma," Astronomy, February 1993, page 38.
9. "Violence in the Neighborhood," Astronomy, November 1993, page 28.
10. "The Galaxy Time Machine," Astronomy, March 1995, page 44.
11. "Quasars: Fires at Cosmic Dawn," Astronomy, August 1995, page 36.
12. "Gamma Ray Burstars: Near or Far?," Astronomy, December 1995, page 56.
13. "Our Strange, Scrappy Ancestors," Astronomy, December 1995, page 52.

DEMONSTRATIONS AND OBSERVATIONS

Modeling the Active Milky Way

To consider how our Milky Way might have looked in its more violent past, as an extended radio galaxy, use the plastic model already mentioned and two large beach balls, maybe 2-3 feet across. Hold the model with the disk on its side; the galactic poles should lie to the left and right of the model. Place one beach ball about five feet to the left, the other about the same distance to the right of the model. If you like, spray paint your beach balls very dark red to remind the students that they are not optically visible, but huge radio emitting clouds of gas now expanded far larger than their home galaxy. Note also that as they cool, they may themselves condense into new, smaller galaxies such as the Magellanic Clouds. If you wanted to model the jets which produce these lobes, use a piece of kite string, sunning through the galaxy's nucleus, up and down to both lobes. Fray it as it gets close to the lobes and expands, much like Figure 17.6; today this string is the Magellanic Stream, the old umbilical cord still hanging in space.

Observing Centaurus A

Of all the AGNs, only this one is optically interesting in small telescopes. There are a few Seyfert galaxies bright enough to be visible, but only an unusually bright nucleus makes them appear any different than a normal spiral visually. Quasars, as their name implies, look just like faint stars in the eyepiece. But NGC 5128 in northern Centaurus is fascinating in even a 6" telescope. Unfortunately it lies far south in the Spring sky and is best seen from latitude 35 degrees N and lower; if you are in the southern U.S., go for it.

At first glance in a 6" at 40x, it seems to be a round E0 galaxy. It is bright enough to spot even in large binoculars. But look more closely--see the broad, dark dust lane occurring right across nucleus, just as in figure 20.8? A normal elliptical galaxy, as noted in chapter 16, should be almost free of gas and dust; this strong radio source has a tremendous amount of dark material silhouetted directly in front of its nucleus. It is an extended radio galaxy; the radio emissions come from dual lobes over a degree above and below the plane of the dusty disk. Additional radio energy comes from a very compact jet, only four light year long, located at the nucleus; this jet is aligned with the much larger radio lobes. Obviously we are looking at the products of two different episodes of activity in this galaxy's nucleus. Perhaps tens of thousands of years ago, the jets spewed out the material that has now expanded into the large radio lobes. The shock wave from this activity blew dust outward from the nucleus, compacting it into the dark lanes. These lanes, when observed in a hydrogen alpha filter, prove to be undergoing an orgy of starbirth, with pink H II regions throughout the dusty disk; stars are forming here at a rate 10x faster than in our own reasonably active Milky Way. Now a new episode of activity, revealed by the tiny jet near the nucleus, is rocking the entire galaxy yet again.

About five years back amateur astronomers spotted a new supernova, just in front of one of the dust lanes. It was easily visible in amateur instruments for several months. It pays to check out the local galaxies--they may be your key to celestial immortality.

While you are observing that far south, drop down just a few more degrees to observe the finest globular cluster in the sky, omega Centauri. It is too far south to see well north of about 33 degrees latitude; if you do have a dark, clear southern horizon, it is easily visible as a fourth magnitude blur with the naked eye and appears as large as the full moon in binoculars.

ANSWERS TO STUDY EXERCISES

1. The spectrum of the radiation is nonthermal in shape, and the light is polarized, as would be expected for synchrotron radiation.

2. The Milky Way shows the same signs of galactic violence as do the more active galaxies, just to a lesser degree. These include ionized gases, nonthermal radiation over a wide range of wavelengths, and high velocity clouds away from the nucleus.

3. Quasars have the largest red shifts of any objects known, indicating they are receding the fastest and by Hubble's Law, are the most distant known objects. Also, the dark lines in their spectra indicate their light has had to pass through cool clouds of intergalactic gas. Finally, in numerous instances quasars have been found associated with far fainter clusters of galaxies, with all objects in the cluster having comparable red shifts and distances.

4. The dark lines, or absorption lines, in the quasars' spectrum must come from their light passing through cool, transparent gas clouds between the galaxies.

5. If quasars really are as distant as their red shift distances imply, then the best model for the tremendous energy production is probably matter falling into a supermassive black hole. Such a huge mass could generate the tremendous gravitational potential energy that is converted into other forms of energy as matter falls inward toward the event horizon.

6. Active galaxies often have two large radio-emitting regions, or lobes, located outside of the galaxy itself. In some cases, the radio-emitting regions coincide with jets of material extending outward from the galaxy.

7. Quasars appear far smaller, almost points of light, and have much larger red shifts than most active galaxies. But in both types synchrotron radiation is typical, as are dual lobes of radio emission and jets coming out of the nuclei.

8. High-speed electrons move close to the speed of light and are focused by a magnetic field. As the electrons move in the field, they emit synchrotron radiation. The rate of this energy loss is greatest at high energy; thus, the X-radiation extends outward only a few light days. To resupply the high energy electrons, a supermassive black hole is the most popular candidate source. Both active galaxies and quasars produce nonthermal radiation from small regions of space, where the intense magnetic fields exist.

9. Were the Hubble constant doubled, then as $D = v/H$, the distances would be halved.

10. As nothing travels faster than light, the region of light variation can't be any larger than the speed of light times that variability, or just a few light weeks across, in some cases.

11. The broad emission lines arise from the motions of the clouds and filaments of hot gas, moving at very high speeds in the strong magnetic fields.

12. The generic model proposes that a supermassive black hole (surrounded by an accretion disk) is eating up material in the nucleus, with jets carrying energy back out to light up the cosmos.

13. Were the bright core hidden, then the generic model proposes we would observe a radio galaxy, with extended radio lobes perpendicular to the disk emitting a lot of radio energy.

14. Were we looking at the source face on, the bright core would overpower everything else and appear as a tiny quasar.

15. The dusty disks absorb most visible and shorter wavelengths, then as the dust grains are heated, they radiate this energy back into space in the infrared as heat.

16. The emission lines tell us the majority of the energy comes from a hot, thin gas.

ANSWERS TO PROBLEMS & ACTIVITIES

1. The classical doppler shift solution gives $v = .16c = 48,000$ km/sec. Using $H = 20$ km/sec/Mly gives $D = 48,000/20 = 2.4$ billion light years. However, 3C 273 is at the practical limit at which this equation can work; any higher recession velocity would require instead the use of relativistic red shifts.

2. If the quasar's total luminosity is 10,000 times our Galaxy's, or 4×10^{37} Joules. Then using $E = mc^2$, we get $m = 4 \times 10^{37}$ J $/ 9 \times 10^{16} = 4 \times 10^{20}$ kg/sec being converted into energy. As the most efficient fusion reaction is only converting .007 of the mass into energy, this implies a total reacting mass of 5.7×10^{22} kilograms of hydrogen being fed into the energy engine every second, or the consumption of a solar mass in about a year.

3. As the Schwartzchild radius requires the escape velocity $= c$, then $R = 2$ G M $/ c^2$, so we get $R = 2 \times 6.67 \times 10-11 \times 10^8 \times 1.9 \times 10^{30}$ kg $/ 9 \times 10^{16} = 2.8 \times 10^{11}$ meters, or about 2 A.U.

4. Obviously the quasar is NOT traveling at 3.53c, so here we must apply the relativistic red shift mentioned in problem 1 above. The actual velocity is found by the relativistic correction $Z = 3.53 = \{(1+v/c)/(1-v/c)\}^{.5} - 1$; thus $4.53^2 = 20.52 = (1+v/c)/(1-v/c)$; solving for v/c, we get $v = (19.52/21.52) \times (300,000$ km/sec$) = 272,000$ km/sec; using $H = 20$ km/sec/Mly, we find the distance to this quasar is 13.6 billion light years.

5. Sin $x = 100,000$ ly $/ 5 \times 10^9$ ly $= 2 \times 10^{-5}$; thus, the Milky Way would subtend about 4 arc seconds at a distance of 5 billion light years.

CHAPTER 21

Cosmic History

CHAPTER OUTLINE

Central Question: How have the physical properties of the universe changed since its origin in the Big Bang?

CHAPTER OVERVIEW

Teacher's Notes

We finally come to physical models of the universe, of physical cosmology. Since covering this chapter is probably separated in time from Chapters 1 through 4 and Chapter 7, it might be good to refresh the students' memories of the observations and gravitational theories to Einstein's gravitational theory and to the observations of the past fifteen years. Keep the treatment of the 3 K background radiation simple. The main point to get across at this level is that the background is strong evidence in favor of the big bang cosmology and that its isotropy indicates a high degree of uniformity early on. Once the big bang model is adopted by the class, the next step is to decide between an open or a closed universe. This question is much harder to resolve, for philosophical reasons, if not for lack of observational accuracy. In deciding between big bang and steady state cosmologies, students need not be influenced by the author's stated bias. But in deciding between an open or closed universe, they may be strongly affected by personal biases. Although some observations do support an open universe, for some reason a closed universe is more acceptable to most people. An oscillating universe gets around the idea of an absolute beginning and an ultimate, irreversible end.

This bias is seen even in the terminology used by astrophysicists or cosmologists, who speak of the "missing mass" problem. The mass we see in the universe is not the mass that should be there-- there's something missing (which many astronomers have been trying to find); the universe should be closed...or should it? you might circumvent problems with the complexity of the details of cosmological models by initiating open discussion of these questions.

Students may get the idea that because of the lack of observations, cosmological models are the result of wild speculations, or of cosmologists "dreaming up models of the universe." Cosmological models, like models of atoms or stars, require extensive calculations consistent with theory to gain validity. While the detail for the theory and calculations are not appropriate for an introductory course, don't underestimate their importance.

One of the most astounding aspects of modern cosmology is that cosmologists have been able to refine the models to the point where the initial expansion of the universe can be described minute by minute. Everyone might not agree on the exact time a particular reaction begins to go in one direction, but it's still somewhat amazing that we can even talk about it.

The inflationary universe provides a wonderful opportunity to discuss elementary particle physics and to show the unity of the physical world in the context of GUTs and cosmology. Point out that they modify the Big Bang model but still keep its basic, and that they are based on general relativity.

It is important to note that all material on the steady-state cosmology has been deleted from this text because the evidence for the Big-Bang model is so strong. Some students will probably have heard of the Steady State model and ask about it. You need to decide whether it is worth discussion.

20.1   Cosmological Assumptions and Observations

"In the beginning" is a fundamental question that intrigues all of us. In this chapter the lines dividing science and religion definitely become fuzzy, and class discussions may get interesting in terms of the Judeo-Christian Genesis and the Big Bang Theory.

The first assumption in the Cosmological Principle (which is really a statement of faith) is that the physical laws work the same everywhere, as Einstein put it, "God's doesn't roll dice". Gravity works the same in pulling the pencil from my hand down to the tabletop as it does in pulling a star into the accretion disk of a quasar's black hole.

The second assumes that while we do have superclusters and voids between them, that on the scale of the entire cosmos, such perturbations are much smaller than the universe itself. Zeilik's mountain analogy is a very good one.

The third one is harder for some students to accept. If all other galaxies are receding from us, as Hubble's Law implies, then are we not stationary at the center of creation? No, we are moving along with everyone else in the expanding fabric of space and time. We, as Einstein's special relativity notes, do not have a privileged position--the overall structure will be similar, regardless of the position or direction of motion of the observer. The analogy of us being riders on one of the dots painted on an expanding balloon being blown up is a good 3D model of what is really happening in all four dimensions of space-time.

The flat geometry was that of the steady state theory, with an infinite but static universe. It is also possible for the ultimate fate of the Big Bang. For the universe to expand just so far, then stop dead in its tracks, would require that the actual density of the universe be exactly equal to its critical density, an unlike coincidence.

The hyperbolic geometry is the ice fate in Frost's classic quandary, "Fire or Ice". If there is not sufficient mass to finally stop the expansion, the open model of the Big Bang will triumph; the universe will use up its hydrogen, with the last red dwarfs fading away trillions of years in the future. The amount of matter presently known in the form of luminous material is less than this critical density, so this model seems favored at present.

The spherical geometry implies closure in the closed Big Bang. The dark matter is out there, in an abundance great enough to exceed the critical density. Gravity will win out; perhaps some 40 billion years down the road, the value of the Hubble Constant, after dropping as the expansion slows, will become negative. The distant galaxies will then begin blue shifting, and the universe will undergo the "Big Crunch", to fall back into a fiery finish in Frost's poem. I personally favor this model, for the cosmos seems to run in cycles, perhaps even this ultimate one with its promise of an oscillating model for the Big Bang, with cosmic rebirth, deja vu?

20.2 The Big Bang Model

Consider the evolution of just the most common element, hydrogen. As discussed earlier in chapter 15, if the hydrogen is over 3,000 K, it is ionized and glows bright pink as an H II region; once the temperature drops below that critical point, the electrons attach themselves to their proton nuclei in stable orbits. This neutral hydrogen, or H I, is now transparent to radiation.

The energy of the expanding Big Bang had been carried out in the expanding red fireball prior to the temperature dropping to about 3,000 K. Now it was allowed to escape into space as the hydrogen suddenly faded into a cosmic H I region. In the universe now over 10,000 times older, larger, and 1,000 times colder, we can still detect this tremendously red shifted energy as the cosmic background radiation, an echo of the Big Bang. Analysis of the COBE data showed the temperature to be 2.73°K, and with just enough large scale variations to led to the condensations of superclusters of galaxies, a relic of the inflationary era.

21.3 The Cosmic Background Radiation

While Wilson and Penzias accidentally found the microwave background in 1964, a group at Princeton had in fact already predicted its existence. When the two groups did get their theory and observations together, it was one of the magic moments in science.

While the first COBE results showed the overall radiation background to be very smooth, more detailed recent studies in late 1992 showed the variations on a small scale that were to lead to the formations of superclusters and voids, in keeping with our mapping of galaxy distribution.

21.4 The Primeval Fireball

As already noted, the Big Bang did not happen at the "center" of the universe; it WAS the entire universe that was so created, with the dimensions of space time growing ever since.

In considering the various eras, remember Einstein's famous $E = mc^2$. The higher the temperature and energy, the more massive the particle-antiparticle pair that would be created. Thus the heaviest particles, neutrons and protons, were created by gamma ray annihilation first, at temperatures of about ten trillion K. Exactly what happened to the antiprotons and antineutrons which should also have been created is still a major mystery. If the symmetry were exact, then seemingly every matter-antimatter pair would have collided, to be annihilated and turned into energy again. This symmetry was not perfect--you are living proof of the triumph of matter. Where the antimatter went is one of those Nobel Prize questions.

At temperatures 1,800x cooler, and perhaps 3 seconds after the Big Bang began expanding, the electron-positron pairs crystallized out of the energy mix, again with the electrons persisting to the present. Note that there were a host of other massive particles and their antiparticles created. The others were unstable and have long since decayed. Even the neutron will decay in about a 1000 seconds, if not stabilized by the strong force in an atomic nucleus. One reason we wanted to build the (now canceled) supercollider in Texas is to push the energies of the collisions up to the point of recreating the conditions of the early Big Bang, to study those first fractions of a second.

I have real problems with the term "recombination;" think about it--if in the hot early cosmos the electrons and protons were always ionized and separated, just how could they "recombine," in the production of the cosmic background radiation. This had to be the FIRST time that an electron could stably orbit a nucleus. The term decoupling, referring to the subsequent separation of now transparent matter from energy, is a better one to me.

Again, the most recently released COBE data does show such small perturbations in the background radiation on exactly the expected scale to be the seeds of the first galaxies Zeilik shows. So far, the COBE data has dramatically supported the inflationary version of the Big Bang Model. I can remember several supermarket tabloids that had headlines in 1992 that scientists had disproved the Big Bang, or COBE has photographed God's eye. To the contrary, COBE has found the cosmos much as we had predicted it, based on the Big Bang as Genesis.

21.5 The End of Time?

To grasp the critical density which determines our fate better, consider just how dense we are; some of us are more dense than others, as the final exam will reveal. You and your students are made chiefly of water, with a bulk density defined as 1 gram/cc; we thought we were dealing with immense densities in white dwarfs (10^6 g/cc) and neutron stars (10^{12} g/cc).

The critical density is about 5×10^{-30} grams/cc. So you are about 2×10^{29} times denser than the universe as a whole; we are the result of a great deal of clumping that turned a hot vacuum into a clumpy cosmos of superclusters, galaxies, stars, and planets.

21.6   *From the Big Bang to Galaxies*

The newest results of the COBE microwave mapping show that such clumping was indeed present very early on. This in turn strongly supports the inflationary revision of the Big Bang Theory. I remember reading the headlines in the supermarket tabloids that COBE had proven the Big Bang wrong....quite the contrary. The Big Bang's name was challenged by the contest in <u>Sky & Telescope</u>. Finally the judges decided no entry better fit the observations, and the Big Bang won yet again.

21.7   *Elementary particles and the Cosmos*

As both gravity and the electromagnetic force obey an inverse square relation between intensity and distance, Einstein believed they were connected in some Grand Unified Field theory; he did not succeed in formulating it, although he worked on it for about 30 years, and he had left notes about it beside his death bed. But since 1954, we have made some progress on the GUTs.

In considering the four forces, remind students that gravity pulls massive objects together. The electromagnetic force is vital to all chemical changes, since they occur in the electron orbitals and depend on the attractions and repulsion of charges and magnetic fields. The strong nuclear force explains how the light atoms can fuse to make heavier nuclei, while the weak nuclear force explains radioactive decay, such as the U 238 decaying in our core now.

For my students, I think it is sufficient that they remember just the order of separation. The gravitational force gained a separate identity first, at the hottest temperatures. The expanding universe next had the strong force appear, then the electromagnetic and weak forces gained separate identities last. The inflationary model depends upon the special conditions associated with the separation of the strong and electroweak forces at about 10^{-35} sec. In this epoch, a light excess of matter (one part per billion) created the triumph of matter over antimatter we see around us as the universe today. It is believed the clumpiness in the COBE data in figure 18.13 stems from that era as well.

21.8   *The Inflationary Universe*

The inflationary model explains why our observations of the average density of the cosmos come close the matching the estimated critical density by noting that with rapid expansion early on (the inflationary epoch), irregularities would be smoothed out, much like blowing up a balloon and watching the stretched skin grow smoother. It also solves the horizon problem by making the initial Big Bang much smaller than previously thought, then having it rapidly expand, thus preserving the uniform background temperature observed in the cosmic background radiation.

As to the end of the cosmos, the search for dark matter continues. If the distant galaxies and quasars are slowing down enough, if H is in fact decaying over time, then as Zeilik notes, we might hope such proof would come from Hubble Space Telescope or Keck data. In our lifetimes, we may learn the ultimate fate of the universe at last. A good overview of the current state of affairs is "Cosmology: All Sewn Up or Coming Apart at the Seems?," by Roth and Primack in <u>Sky and Telescope</u> for January 1996 on page 20.

ADDITIONAL RESOURCES

Slides 21.13b and c and transparencies 21.2ab, 21.3ab, 21.4, 21.7, 21.9, 21.11, and 21.16 will enrich your class presentations.

Two good videos fit well with this chapter. The third episode of The Astronomers series, "A Window to Creation", focuses on the discovery of the echo of the Big Bang. Timothy Ferris' award winning "Creation of the Universe" is 90 minutes long, but widely available in video stores; you might recommend your students check it out. Both can be obtained from the Astronomical Society of the Pacific; call (415) 337-2624 for their catalog.

SUPPLEMENTARY ARTICLES

1. "Matter and Evolution of the Universe," Astronomy, September 1984, page 67.
2. "The Fate of the Universe," Astronomy, January 1986, page 6.
3. "To the Big Bang and Beyond," Astronomy, May 1987, page 90.
4. "The Cradle of Creation," Astronomy, February 1988, page 40.
5. "The Search for Dark Matter," Astronomy, March 1988, Page 18.
6. "The Structure of the Visible Universe," Astronomy, April 1988, page 42.
7. "Recreating the Universe," Astronomy, May 1988, page 42.
8. "Supercomputing the Universe," Astronomy, December 1989, page 48.
9. "Is the Universe too Smooth?," Astronomy, June 1990, page 20.
10. "Is Cosmology a Sometime Thing?," Astronomy, July 1991, page 38.
11. "Shedding Light on Dark Matter," Astronomy, February 1992, page 44.
12. "Beyond the Big Bang," Astronomy, April 1992, page 30.
13. "Wormholes: Tunnels Through Time," Astronomy, June 1992, page 28.
14. "COBE's Big Bang -- How Galaxies Began," Astronomy, August 1992, page 42.
15. "Counting to the Edge of the Universe," Astronomy, April 1993, page 38.
16. "A New Map of the Universe," Astronomy, April 1993, page 44.
17. "How Old is the Universe?," Astronomy, April 1993, page 38.
18. "The Great Attractor," Astronomy, June 1993, page 38.
19. "In the Beginning," Astronomy, October 1993, page 40.
20. "Everything You Wanted to Know About the Big Bang," Astronomy, January 1994, page 28.
21. "How Old is the Universe?," Astronomy, October 1995, page 42.
22. "Curtains at the Edge of the Universe," Astronomy, November 1995, page 48.
23. "Globulars and the Age of the Universe," Astronomy, February 1996, page 44.
24. "Wormholes and the Spacetime Paradoxes," Astronomy, February 1996, page 52.
25. "A Window into the Deep," Astronomy, April 1996, page 82.
26. "What Happened Before the Big Bang?," Astronomy, May 1996, page 34.
27. "A River in the Universe," Astronomy, August 1996, page 44.

ADDITIONAL DISCUSSION TOPICS
Cosmic Nucleosynthesis

From the dense conditions of the first minutes, cosmic nucleosynthesis produced the material we will still find in the oldest population II stars. Neutrons and protons collided directly, to produce deuterium or heavy hydrogen nuclei. Two deuterium nuclei in turn collide to make helium 4 directly, a far faster process than the three collisions in the proton-proton cycle inside main sequence stars, even though the basic reactants (four protons) and product (a He 4 nucleus) are the same. But as the free neutrons decayed into proton-electron pairs, no more deuterium could be make, so production of helium stopped after about 30 minutes. It is amazing just how efficient the conversion of H 2 into He 4 was; about 25 % of the known mass of the universe is believed to be helium, yet deuterium on the earth makes up only about 1 out of every 7,000 hydrogen atoms in seawater (deuterium oxide is "heavy water"). Some cosmologists use this abundance of deuterium to argue that the universe will not be dense enough for closure.

So what about the elements heavier than helium, the chief ones that make up the dense terrestrial planets? They were to be born much later, forged in the cores of highly evolved giant stars. Remember that the Big Bang is cooling as the rapidly expands. But back in chapter 13, we noted that while hydrogen fusion took temperatures of about ten million degrees, the triple alpha process required far higher temperatures, about 100 million K, to work. Thus by the time the Big Bang built enough helium 4 to do anything with, the cooling temperatures stopped it from doing it. The Big Bang thus made almost nothing but hydrogen and helium, which are still the main materials in most stars, and still comprise about 99 % of the observed universe. Ours is still a young universe, with plenty of hydrogen and helium to play with. Of course with time, subsequent generations of supernova will enrich the interstellar medium with more and more heavy elements, making planets like earth easier to make and even more common, perhaps a reassuring finding for our descendants if they decide to go out into the galaxy.

Olber's Paradox

A topic worth noting is just why the night sky is dark. In Olber's Paradox, we consider an infinite universe (the flat geometry model), filled with an infinity of stars. Thus in every direction we look, we should be staring directly at the surface of another star. The entire sky should be as bright as the photosphere of the sun itself.

The Big Bang explains the darkness of the real night sky in two ways. First, if it only began 20 billion years ago, the universe is not nearly old enough, nor will it contain nearly enough stars to light up the night sky as Olber's paradox called for.

Secondly, remember the great red shifts of the quasars and distant galaxies. As the universe expands, these red shifts dilute the energy being radiated by such objects. A photon of ultraviolet light was shifted all the way into the infrared in our example quasar in the last chapter. As infrared is far less energetic than ultraviolet, the distant objects appear much less bright to us than if they were in a static universe.

The Hubble Time and Radius

Another point worth making is the Hubble Time. Consider the nature of Hubble's constant. It is typically H = 15 kilometers/second/Million light years. But what do both kilometers and light years measure? Distance, of course. If we were to invert the Hubble Constant, H, and substitute in the number of kilometers in a million light years, we would be left with the age of the universe in seconds; it comes about to be about 20 billion years old, if we use this small value in the range of H. If we favored a universe expanding more rapidly, with H about 30, then 1/H, the Hubble Time, would be halved to just over 10 billion years. It makes sense; the faster the universe is expanding, the less time it will take to run out cosmic movie projector backwards to creation.

Another way of considering Sagan's "edge of forever" is the Hubble radius; just how far away from us can anything be that we could observe. As Einstein says that nothing can travel faster than c, the speed of light, if we substitute c into Hubble's law, we get D = c/H, or with an H of 15, the Hubble radius is 300,000 km/sec / 15 km/sec/million light years, or a radius of 20 billion light years out to this edge; note this in exact agreement with the Hubble time for the age of the universe with the same value of H.

The Arrow of Time

An interesting aside to the Closed model is discussed by Sagan in Cosmos 110, "The Edge of Forever." While we are used to reversibility in three dimensions (I can move left or right, backwards or forwards, up or down), time is not now so blessed. When we wake up, it is always tomorrow, never yesterday. The arrow of time flows toward tomorrow because the expanding universe is larger tomorrow than it is today. If gravity wins, and the universe undergoes a collapse billions of years from now, would we witness causality reversal, with time flowing backwards. In such a universe, the students would begin with the final exam, and end on the first day of class (perhaps not a bad option for some, at this point in the course).

DEMONSTRATIONS

Get a good sized round balloon, of fairly durable material, either clear or light in color. A 15-20" inflated size will be adequate for classroom demonstration. Paint on it with some water soluble markers a host of dark dots, to represent galaxies; if your use alcohol or other solvent based markers, they may weaken the balloons surface in time and create an embarrassing version of the Big Crunch model in front of your class. Before you begin inflating the balloon, have a student measure the separation distance between any two dots on the surface with a tape measure. After it is well inflated, use a tape again (carefully) and note how the distances between our model galaxies have increased in our open model for the Big Bang.

Not everyone is so sure the Big Bang is that uniform. Some think that like a balloon with a poorly molded surface, some areas (the voids) expand faster, while others (the supercluster regions) lag behind; perhaps your balloon may be this realistic.

ANSWERS TO STUDY EXERCISES

1. The Big Bang model requires the universe to look the same to all observers (the Cosmological Principle). Also, it assumes the universe is isotropic, homogeneous, and obeys the same physical laws at all places at all times (the laws are universal).

2. Some fundamental cosmological observations include:
a. The night sky is dark.
b. Elemental spectral lines from distant objects appear like the elements of the earth and sun.
c. Laws of motion apply to stars and galaxies and clusters of galaxies just as they work here.
d. the universe appear the same regardless of the direction you look (clumping of matter is not a local phenomenon).
e. Distant objects all appear to be receding from us, the farther ones receding faster.
f. We detect an isotropic background microwave background radiation at 2.73 K.
g. This microwave radiation shows small-scale fluctuations.

3. Observations b and c allow us to interpret any observations based on physical laws we understand them based on local experiments. Observation a tells us the universe is evolving over time, and e gives one of these changes is the expansion of the universe. Observation d implies the expansion is isotropic, f suggests that the expansion began from a hot, compact state, and g tells us the first units to form were probably the superclusters of galaxies.

4. COBE shows the microwave data is a very close fit to a blackbody curve at 2.73 K; this low temperature suggests a cosmic origin for the radiation, since any other source would probably be considerably warmer. The isotropy of the radiation eliminates local sources, such as the sun, solar system (not in ecliptic plane), and Galaxy (nor in Galactic Plane). Even the fluctuations are on the scale predicted by the inflationary model of the Big Bang.

5. The microwave background supports the Big Bang theory, because this model predicted that a background radiation of low temperature should permeate the cosmos, a relic of the radiation released at the recombination of hydrogen atoms.

6. The Big Bang made chiefly hydrogen and helium, with traces of deuterium, tritium, lithium, and beryllium. Heavier elements like those which make up the terrestrial planets did not have a chance to form, for they require both abundant helium and high temperatures in the triple alpha reaction to be formed. By the time the helium concentration had built up, the expanding cosmos had cooled down to the point such production was no longer possible. Instead these heavier elements (carbon and beyond) had to be made later in the cores of massive stars.

7. The supporting evidence includes the microwave background radiation, the observed recession of distant objects (Hubble's Law), the estimated ages for the oldest globular clusters, and the observed helium abundance in the oldest celestial objects.

8. Just as energy can be formed from mass in fusion reactions, so collisions between very energetic photons of energy can form particle-antiparticle pairs; this conversion of course is the essence of Einstein's $E = mc^2$. However, it took a lot of energy to form any massive particle; the high temperatures and energies needed for cosmic nucleosynthesis persisted for only the first seconds of expanding and cooling Big Bang.

219

9. Ours is a cosmos dominated by matter, yet the model predicted that just as much antimatter was made as normal matter. Where has the antimatter gone? The inflationary revision suggests that the imbalance between the two froze out early on, making ordinary matter the dominant form in the observable portion of the universe.

10. If the cosmos were flat in spacetime geometry, then the universe must have been very, very close to flat at the time of the Big Bang. The inflationary model copes by expanding spacetime so fast that small deviations (recently confirmed by COBE) from perfect flatness will be smoothed out.

11. The peak for the microwave data falls at 2.73 K, in good accord with the prediction of the Big Bang that the decoupling of matter and energy as the cooling hydrogen turned from a glowing H II region into a cooler, transparent H I cosmos. Based on the event happening at about 3,000 K, the present cosmos must be about 1,000 older, colder, and larger than the cosmos at that moment.

12. The observed fluctuations were the seeds for the formation of the superclusters of galaxies.

13. If the estimated 90% of the universe is indeed dark matter, that will prove enough mass and density for closure, favoring the closed (or oscillating) version of the Big Bang. If less than this critical density of matter is out there, the expansion will slow, but never grind to a halt, thus giving an open Big Bang to expand and cool off forever.

14. As we go back in time, the temperature was higher, until about a million years after the Big Bang we find in the 3,000 K fireball the hydrogen recombination event. To find the approximate temperature of the fireball in this time interval, the further forward in spacetime we go, the colder the fireball becomes; at two million years, the fireball was only 1,500 K, for instance.

ANSWERS TO PROBLEMS & ACTIVITIES

1. Wien's law, again; peak wavelength = 2.9×10^{-3} / 2.73 K = .0011 m = 1.1 millimeters.

2. $E = hf = 6.67 \times 10^{-34}$ / 10^{22} = 6.7×10^{-12} J; as $E = mc^2$, thus m = 6.7×10^{-12} / 9×10^{16} = 7.4×10^{-29} kg. These gamma rays could have been mesons and leptons such as electrons, but nothing as heavy as the hadrons like protons and neutrons, which are around 1.6×10^{-27} kg.

3. The energy equivalent of the electron mass (9.1×10^{-31} kg) is $E = mc^2 = 9.1 \times 9 \times 10^{-15}$ = 8.2×10^{-14} J. As $E = h\,c$/wavelength, then this photon's wavelength can be calculated to be $6.67 \times 10^{-34} \times 3 \times 10^8$ / 8.2×10^{-14} = 2.44×10^{-12} meters, in the gamma ray portion of the spectrum. As the proton is about 1,836 times as massive, it would require 1,836 times more energy or be produced by a gamma ray of only 1.33×10^{-15} meters in wavelength.

4. At 3,000 K, Wien's Law tells us the peak wavelength = 2.9×10^{-3} / 3,000 K = 9.67×10^7 m or 9,667 Angstroms, in the infrared portion; this is about the surface temperature of Antares or Betelgeuse, so it would have appeared orange red in the visual range. Today this peak is at 1.1 millimeters, in the microwave portion of the spectrum, and entirely invisible to our eyes.

CHAPTER 22

Bios and Cosmos

CHAPTER OUTLINE

Central Question: What are the characteristics and possible origin of life as we know it, and what consequences follow from these for the possibility of life elsewhere in our Galaxy?

22.1   The Nature of Life on the Earth
    A. Organisms
    B. Proteins and nucleic acids
    C. Clues from astronomy
    D. The spark of life
    E. Synthesis of simple organic molecules
    F. Synthesis of complex molecules

22.2   The Genesis of Life on the Earth
    A. Clues from biology
    B. Clues from geology

22.3   The Solar System as an Abode of Life
    A. Mars: the best chance
    B. Amino acids in meteorites
    C. Mass extinctions on earth?

22.4   The Milky Way as an abode of Life
    A. Cosmic prospecting
    B. Astronomical factors
    C. Biological factors
    D. Speculative sociological factors
    E. The numbers game

22.5   Neighboring Solar Systems?
    A. Center-of-mass motions
    B. Doppler shift detections
    C. Other evidence

22.6   Where are They?
      A. How far to our galactic neighbors?
      B. Are we alone?
      C. The search for extraterrestrial intelligence

22.7   The Future of Humankind
      A. Growth
      B. Space colonization
      C. Beyond the solar system
      D. The future of technological civilizations

Enrichment Focus 22.1: Modeling Stellar Ecospheres

CHAPTER OVERVIEW

Teacher's Notes

In discussions of the possibilities of life elsewhere in the universe, questions that students often raise are generally concerned with four aspects of the problem: (1) the definition of life, (2) the uncertainties in the probability estimates (numbers game), (3) communication with other civilizations, and (4) interstellar travel (to and from earth).

If the rule of thumb given at the beginning of the chapter for identifying living things is not sufficient for you, then build a definition on the common chemical basis for terrestrial life--this defines "life as we know it" or LAWKI. If students suggest other life forms, point out that one fact about life as we know it is that we don't know of any other. To let imaginations run wild, with no observable evidence for other life forms, makes discussion of probabilities meaningless. A popular science fiction treatment of other life forms is Michael Crichton's The Andromeda Strain (Alfred Knopf, 1969). One point made in that novel is our inability to recognize other life forms. For example, if a rock were "living", but of a form that carried out its life processes on times much longer than a human lifetime, we would have trouble recognizing those life processes.

Perhaps the best (but dated) reference for the details of numerical estimates of numbers of life-supporting stars is Sagan and Shklovskii, Intelligent Life in the Universe (Delta, 1966). Of course, estimates will vary from one text to another, depending strongly on the author's personal biases. It might be interesting to poll your class before reading or discussing this chapter as to their preconceived notions and opinions about extra-terrestrial life. Use questions that are easily tallied, such as having simple yes/no answers. Then do the same survey after finishing this chapter; let the class know the results of the two surveys.

When playing with the numbers themselves, point out the probabilities, and resulting numbers of stars supporting intelligent life, are considered in this text only for the Milky Way Galaxy. The numbers increase tremendously when the entire universe is considered. For instance, in an extreme case, assume we are unique to the Milky Way but typical for the universe; that is, the frequency of life-supporting planets is one per galaxy (a well established lower limit from local observations). Still, the number of similar galaxies is huge for the universe as a whole. It's easy to reach the conclusion we are alone due to the vast distances.

Here are arguments you can use to establish some of the numbers more firmly. A factor that may influence estimates of the number of stars with planetary systems is the rotation speed (or angular momentum) of different stars. (Students should be familiar with this term by now--recall the angular momentum problem from Chapter 13). Some astronomers believe that a slow rotation of a star signifies that planets orbit the star, because much of the angular momentum of the original gas cloud would be transferred to planetary motions when the star forms. Most solar-type main sequence stars are slow rotators.

Another factor is the possible existence of a physical basis for the Titus-Bode Law. if such exists, then the probability of planets forming within the life zones (or ecospheres) of solar type stars are very large. The new solar systems found in 1996 do not conform to the pattern we find here; often massive planets lie far closer to their stars than do our jovians to the sun.

There have been several attempts to detect signals from extraterrestrial civilizations. Project Ozma, the first such search, is mentioned only briefly in the text. Some of the later attempts included transmitting signals toward stars in the solar neighborhood. So far all attempts have apparently been unsuccessful, although in 1993 the Planetary Society reported its META project had found about 200 puzzling signals for further analysis. However, the number of stars examined so far is less than 0.1% of the number required for a reasonable statistic chance of detecting an extraterrestrial civilization.

The distances involved are the biggest obstacles to interstellar travel or communication. An order of magnitude calculation will illustrate this. Using 100,000 ly for the Galaxy's diameter, and 5,000 ly for its thickness, and assuming ten billion stars in the Galaxy, the stellar density is one star for every 400 cubic light years, or an average spacing of about seven light years. If there are as many as ten million stars supporting intelligent life, then the average spacing between civilizations is about 150 light years. For only a thousand lively stars, the spacing increases to over 3,000 ly. Even if the probability for extraterrestrial life is large, we are alone in the vastness of space.

So far radio searches for interstellar amino acids have failed to detect any. A chief effect of the discovery of amino acids would be to increase the probability of extraterrestrial life. Although the materials for building amino acids are known to be present in the giant molecular clouds, one uncertainty is the natural processes for assembling them into the complex molecules that we use to define life. Also, there would be the possibility that the amino acids result from life in the universe, rather than representing one step on the way to building complex compounds.

22.1 The Nature of Life on the Earth

As noted back in Chapter 17, supernovas might have played a role in adding to the genetic diversity by increasing the rate of mutations. The lunar tides also might have been vital, as discussed in chapter 9, in mixing the amino acids in the tidal pools and creating molecules complex enough to become self-replicating.

22.2 The Genesis of Life on Earth

It is amazing just how soon after the earth was cool enough for liquid water oceans that those oceans became populated. No wonder Hoyle suggests that life had a head start, arising on the surfaces of comets (chapter 12) before it ever arrived at earth. The Allan Hills meteorite and its apparent micro fossils similar to earliest earth fossils even suggests life began on Mars first, then migrated to earth via meteorites perhaps about 3.7 billion years ago. Maybe we are all martians!

Certainly carbonaceous chondrites do contain complex amino acids, perhaps formed out in space before the earth and solar system were even born.

Remember that not only did the atmosphere play a vital role in shaping life, but that since it arose, life plays a role in changing the atmosphere. Again, our present absence of carbon dioxide (chapter four) is due largely to carbonate rock deposition by reef builders and photosynthesis enriching our atmosphere with oxygen. Were we to find another planet with such an atmosphere, we would be immediately suspicious that some form of plant life is present there.

22.3   The Solar System as an Abode of Life

The Mars exploration program certainly got a boost from the August announcement of possible micro fossils in the Allan Hills meteorite. For a more complete report, see "Life from ancient Mars," by J. Kelly Beatty in the October 1996 issue of Sky and Telescope. It is likely a host of robot rovers, such as described in the articles on NASA's plans in the solar system unit.

When Galileo flies past Europa in late 1996, close-up photos of the surface of Europa may tell us how thick the ice is and how good the chances are for photosynthetic organism there.

The carbonaceous chondrite material again shows how abundant organic type compounds are universally. The note about left vs. right handed proteins is a good one; we might not be able to digest their food, even if it looked and tasted like our own. Practical use will be made of this for dieters with the development of artificial sugars, fats, and proteins that cook, look, and taste just like the calorie rich ones we crave but are undigestible due to their reversed layout.

Concerning mass extinctions, we have already discussed numerous examples of such impacts, such as the Yucatan impact of 63 million years ago that ended the reign of the dinosaurs. With the recovery of Comet Swift-Tuttle (chapter 12) came the unsettling possibility that we might have something to really worry about by August of 2126 AD; later analysis proved the danger minimal, but we are still constantly sweeping the skies for earth-grazing asteroids, all the better to know their orbits and the potential dangers they pose.

22.4   The Milky Way as an Abode of Life

If the current trend toward nuclear disarmament and international cooperation continues, then human civilizations' reign on the earth may be much longer than many of us expected just a decade ago. Obviously the longer a civilization can refrain from committing suicide, the better a chance there is for us to receive their radio signals and establish contact at last.

22.5   Neighboring Solar Systems?

As reported in "Other Stars, Other Planets," in August 1996 issue of Sky and Telescope on page 20, we are finding several nearby stars have planetary systems. Most do not resemble our solar system, for the first planets found must be massive ones, and the closer they are to their star, the easier we can spot doppler shifts. Still, the more abundant all planets are, the better the chance that at least one of them will possess earthlike conditions necessary for life to arise and evolve.

A problem is that most stars are binaries. For widely separated binaries (like alpha Centauri), terrestrial planets could orbit BOTH stars and double the chances for life. But in closer systems, tidal forces and changing exposures might stop any planets from forming, or make climatic conditions on terrestrial planets too extreme for life to evolve at all.

22.6 Where are They?

I have mentioned several limitations to why life might not evolve elsewhere, or as rapidly as it has here on earth. Again, the sun may be an unusually stable star; most stars may be too variable to give life a foothold. The Milky Way may have passed through a quasar phase just before the solar system formed, thus sterilizing itself of any previous life forms throughout the entire galaxy. See "Are We Alone in the Universe," in July 1996 Astronomy on page 36 for more discussion topics.

22.7 The Future of Humankind

The discovery of even one other intelligent life form elsewhere would be encouraging. It would prove intelligence is a nice thing for living things to possess! To that end, NASA in 1992 began project SETI, with huge radio dishes listening to over 800 nearby stars in hopes of picking up intelligent whispers across the void. But in November 1993, some of the less intelligent life in Washington canceled this ambitious project. Luckily private funding has since picked up the slack, and the finding of apparent martian micro fossils may make future research much more popular.

ADDITIONAL RESOURCES

Transparencies 22.11 and 22.12 will be helpful in class. The second episode of Cosmos, "One Voice in the Cosmic Fugue," is very highly recommended, even if a little dated.

The privately funded META project of the Planetary Society is still seeking first contact. Read their journal, The Planetary Report, for updates. Call them at (818) 793-5100 or write their office at 65 North Catalina Avenue, Pasadena, CA 91106.

SUPPLEMENTARY ARTICLES

1. "Digging Deeper into Life on Mars," Astronomy, April 1988, page 6.
2. "Birth of Planet Earth," Astronomy, June 1989, page 24.
3. "Searching for the Waters of Mars," Astronomy, August 1989, page 20.
4. "What would the Earth be like without the Moon?," Astronomy, February 1991, page 48.
5. "Life Near the Center of the Galaxy," Astronomy, April 1991, page 46.
6. "Does Alpha Centauri have Intelligent Life?," Astronomy, April 1991, page 28.
7. "Demise of the Dinosaurs -- A Mystery Solved?," Astronomy, July 1991, page 30.
8. "Asteroid Impact: The end of Civilization?," Astronomy, September 1991, page 50.
9. "The Earth's Atmosphere: Terrestrial or Extraterrestrial?," Astronomy, January 1992, page 38.
10. "Life Around a Larger Sun," Astronomy, May 1992, page 50.
11. "Lost and Found: Pulsar Planets," Astronomy, June 1992, page 36.
12. "Desperately Seeking Jupiters," Astronomy, July 1992, page 42.
13. "Listening for Life--NASA's SETI Project," Astronomy, October 1992, page 26.
14. "Life on a Metal-Poor Earth," Astronomy, October 1992, page 40.
15. "The Cosmic Origins of Life--Did Comets Help?," Astronomy, November 1992, page 28.
16. "Did Mars Once Have Martians?," Astronomy, September 1993, page 26.
17. "A Friend for Life: Habitable Planets," Astronomy, June 1995, page 46.
18. "Two New Solar Systems," Astronomy, April 1996, page 50.
19. "Are We Alone in the Universe," Astronomy, July 1996, page 36.

ANSWERS TO STUDY EXERCISES

1. Carbon is manufactured in the initial stage of giant star evolution in the triple alpha process. At about 100 million K, three helium 4 atoms are fused into C 12. It escapes these giant cores as stellar winds shed by the evolved red giants, or in their supernova remnants. It then mixes with other atoms to enrich the interstellar medium in life-giving elements, some of which even form organic type molecules in the giant molecular clouds like that near the Orion Nebula. It is from such dense regions that new solar systems, with the dust for terrestrial planets, form.

2. Supernovas liberate the most common elements of earth, such as carbon, oxygen, and silicon, formed earlier in the cores of evolved giants. They also synthesize the rare elements beyond iron in weight. Without the resurrecting force of the supernova explosion, most heavier elements would go to the grave with the stars that gave them birth.

3. Neglecting earth, the best sites are probably Mars and Europa. The negative results of the Viking landers lets us know that at least two sites on Mars are barren. We will have to wait for Galileo's high resolution photos of the thin, icy crust to assess the chances for life in the oceans of Europa. The cold temperatures of Titan make life there unlikely (or at least very slow), despite the abundances of organic molecules in its atmosphere and probably on its surface. It has been suggested the atmospheres of Jupiter and Saturn might have exotic floating forms of life, but again we must wait for more data from Galileo. Even the watery mantles of Uranus and Neptune have some promise, although exactly how you could find out is still uncertain.

4. If life elsewhere evolves at about the same pace as it did on earth, then the very short life spans of hot O and B stars mean that they will not remain stable long enough for life to evolve much before it is exterminated in the ashes of their supernova suicides, in perhaps only 10 million years since the stars reached the main sequence. As MS stars they are far less stability than our sun; such sudden changes would make it very hard for life to evolve and adapt.

5. There are many possible criticisms, such as overestimating the stability of most stars (the sun, a single star, may be far better behaved than most other stars in their complex and volatile families. The planet Jupiter has been suggested as a shield that protects the entire inner solar system from even more frequent comet impacts. And of course the recent end of the cold war give us pause to hope that mankind's future on earth and in the universe may last longer than it appeared back in October of 1962.

6. The giant stars formed the critical elements of life (carbon, oxygen, iron, sulfur, nitrogen, and so forth) in their later stages, shedding some of these elements out in stellar winds, then showering still more elements out in their supernova deaths. In this sense, we are indeed the children of stars that had to die in order that we could be born.

7. The number of new planets forming capable of bearing life will also decline. But as the concentration of heavier elements increases with time, even more terrestrial type planets will be

226

8. Mars seemed promising, for it had liquid water once upon its surface, and apparent microfossils in the Allan Hills meteorite. Yet if any life formed in those waters, it does not seem to have been able to adapt to the drying and cooling conditions as Mars evolved, according to the Viking data.

9. The microwave region includes the "water hole", a good region to call attention to the basic substance of life. Also, this region has relatively little man-made interference as of yet, so it would be easier to pick up extraterrestrial signals there.

ANSWERS TO PROBLEMS & ACTIVITIES

1. The center of mass relation is just the law of levers, so $M_sD_s = M_jD_j$, where the total distance is the sum of $D_s + D_j = 5.2$ AU; $D_j/D_s = M_s/M_j = 1000$. Thus the sun is about .0052 AU, or about 7.8 x 10^5 km from the center of mass of the system. At a distance of about 250,000 AU to alpha Centauri, this would be a wobble of only sin x = .0052 / 250,000 = 2.1 x 10^{-8}, so the angle would only be about .004", far too tiny to find even with the Hubble Space Telescope.

2. $P^2 = A^3$ for solar system objects, according to Kepler's Third Law. If Nemesis is really out there, its average distance is $(2.6 \times 10^7)^2 = D^3 = 6.76 \times 10^{14}$, so D = 87,764 AU. Since this is 35 % of the way to alpha Centauri, this makes it likely close encounters with other stars would have stolen such a far-ranging companion away from the sun long ago.

3. $N_{ic} = R^*P_rP_eN_eP_lP_iL_{ic}$; I would agree than about 20 new stars are formed in the Milky Way per year, but I think many wide binary pairs can also form planets, so I put this at about .8. Considering the ecosphere problem, I agree the chances of getting an earth-like planet at just the right distance are not good, perhaps close to .1. I think it likely a star would be lucky to get even one planet in its ecosphere, so I set the number of planets as only .5 at best. Considering the abundances of organic type molecules in the interstellar clouds, I agree the building blocks are out there, but for them to have had just the right conditions to grow into DNA is less likely. I think P_l only .1 at best. I am also less certain about the probability of intelligence developing; if frequent cometary impacts cause the biological clock to be reset many times, then this might be only .5 at best. As to the lifespan of a technological culture, I will optimistically go with the writers on Startrek and set the spans at a million years. Thus (and student answers will vary widely, of course) I get $N_{ic} = 20 \times .8 \times .1 \times .5 \times .1 \times .5 \times 10^6 = 40,000$ star systems capable of having intelligent life in the entire Milky Way Galaxy.

4. HD 114762 has a period of 84 days, and as an G V star, should have a mass about the same as the sun's. Plugging the 84 day period into Kepler's Third Law, we get a semimajor axis of $P^2 = D^3 = (.229 \text{ years})^2 = .05244$, hence the semimajor axis is .374 AU (note how similar this orbit is to Mercury's orbit). If we use the observed velocity shift of 500 m/s, we get an orbital circumference of 3.63 x 10^6 km, or dividing by 2 x 3.14, a radius of 578,000 km for the orbit of the visible G V star. This radius is about .00385 AU, so the visible star orbits about a hundred times closer to the center of mass than does its invisible companion. This means the companion must be a hundred times less massive; a mass of .01 solar masses means the brown dwarf is only ten jupiter masses.